G200 Youth Forum 2017 Conference

Dubai, United Arab Emirates
6 - 10 December 2017

ISBN: 978-1-5108-6183-1

Printed from e-media with permission by:

Curran Associates, Inc.
57 Morehouse Lane
Red Hook, NY 12571

Some format issues inherent in the e-media version may also appear in this print version.

Printed by Curran Associates, Inc. (2018)

For permission requests, please contact G200 Association
at the address below.

G200 Association
14 rue du Rhone
Geneva 1204, Switzerland

Phone: +41-22-518-28-88

g20youth@g200youthforum.org

Additional copies of this publication are available from:

Curran Associates, Inc.
57 Morehouse Lane
Red Hook, NY 12571 USA
Phone: 845-758-0400
Fax: 845-758-2633
Email: curran@proceedings.com
Web: www.proceedings.com

G200
ASSOCIATION

2017(5)

CONFERENCE PROCEEDINGS

G200 Youth Forum 2017 Conference
December 6-10, Dubai, UA

INTRODUCTION

The G200 Youth Forum (Dubai, UAE, 6-10 December 2017) was an anual international event organized for young leaders from around 200 countries; and over 100 young leaders, students and academics, representatives of the business world and governments participated in it.

It was the 12th year of the Forum. Previous events were held in:

2006 – G8 Youth Summit – Russia (Saint-Petersburg);
2007 – G8 Youth Summit – Germany (Berlin);
2008 – G8 Youth Summit – Japan (Tokyo);
2009 – G8 Youth Summit – Italy (Milano);
2010 – G20 Youth Summit – Canada (Vancouver);
2011 – G20 Youth Summit – France (Paris);
2012 – G20 Youth Summit – USA (Washington D.C.);
2013 – G20 Youth Forum – Russia (Saint-Petersburg):
2014 – G20 Youth Forum – Germany (Garmisch-Partenkirchen, Bavaria);
2015 – G200 Youth Forum – Germany (Garmisch-Partenkirchen, Bavaria); and
2016 – G200 Youth Forum – Germany (Garmisch-Partenkirchen, Bavaria);
2017 – G200 Youth Forum – UAE (Dubai).

The G200 Youth Forum 2017 had 5 main events which run in tandem with each other:

- G200 Youth Summit – Final Perspectives;
- Conference – Publication of academic articles in the Conference Proceedings;
- International Young Parliamentarians' Debate– with Joint Statement;
- The new established G200 Young Professionals Platform and The G200 Entrepreneurs Platform 2017;
- Joint Sessions - with Joint Sessions reports.

Conference 2017 is an international Conference for representatives from the best Universities in the world who are experts in international relations, law, economics, finance, technology, medicine, education and humanities. During the Conference participants discuss global problems which are on the agenda of the academic communities of different countries in the format of 8 round tables working in parallel.

1. Economics and Finance
2. Law and Human Rights
3. World Politics and International Relations
4. Social Affairs and Medicine
5. Ecology, Environment, and Energy
6. Design, Technology, and Innovations
7. Education and Youth
8. Humanities: History, Philosophy, Linguistics, Arts, and Journalis

In total: more than 4700 attendants in 11 years from around 200 countries.

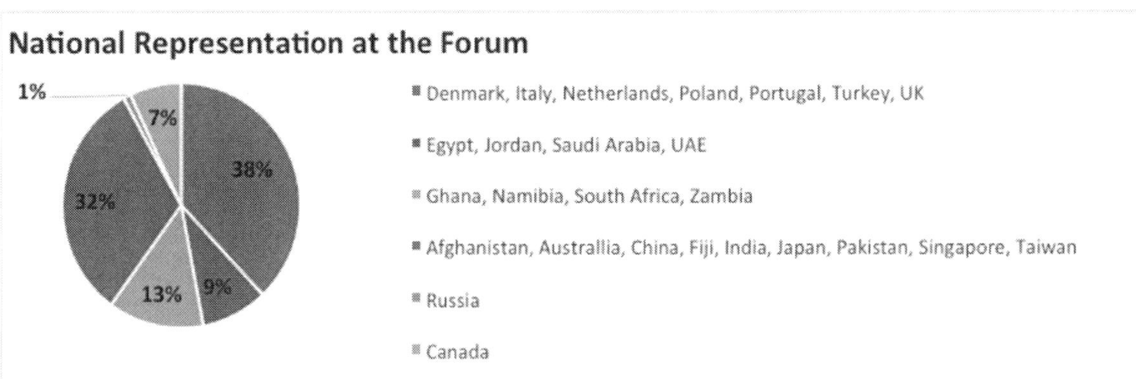

The participants of the Conference were representing the best Universities of their countries:

- Administration of Saint-Petersburg
- Bentley Adolescent Unit: Mental Health Nurse
- Centennial College
- ISCTE - Instituto Universitário de Lisboa
- Leiden University
- Polish Young Academy, Polish Academy of Sciences
- Tianjin University of Finance and Ecoomics
- University of Canberra

Reviewer Acknowledgement for the Conference Proceedings 2017

G200 Association would like to thank all our reviewers who have contributed to the Conference Proceedings 2017

Dr. Zhang Yanling, Professor, Deputy Director of Tourism Department, Business School, Tianjin University of Finance and Economics, Tianjin, China.

Dr. Jagmeet Singh, Assistant Professor, Department of Human Values & Professional Ethics, Punjab Technical University, Punjab, India

In the Conference Proceedings 2017 the spelling and punctuation of the authors are kept. The opinions expressed in the articles are those of the authors and do not necessarily reflect the position of the G200 Association.

CONTENT

Economics and Finance

Cross border e-commerce and its influence on SMEs in China

Li Tang
M.D., Assistant Professor, Tianjin University of Finance and Economics, China

1. INTRODUCTION

With the continuous development of Internet technology, cross-border e-commerce is developing rapidly, and brought considerable convenience for the development of trade, meanwhile, after the financial crisis in 2008, world trade growth started to slow down, in order to open up new markets and increase profits, more and more small and medium sized enterprises in China began to adopt cross-border e-business, expecting to bring new profit increase. According to the data published by China Electronic Commerce Research Center, every year over 90% new business entities registered in cross-border e-commerce platform are SMEs and they played an import part China's cross-border E-business' development, vigorous and fast development of e-commerce has brought new vitality to China's SMEs, but what is the situation of China's SMEs using cross-border e-commerce, What kind of impact does cross-border e-commerce have on SMEs' development, what are the problems and issues China's SMEs facing in developing cross border e-commerce, these are the issues to be discussed in this paper.

2. LITURATURE REVIEW

Development of cross border e-commerce and SME has been discussed in the business and academic circle. But the studies on the topic of development of cross-border e-commerce and its influence on SMEs are relative scarce. Researches are mainly on issues on adoption of e-commerce and factors affecting the adoption of e-commerce of SMEs. Mesut Savrul et al. (2014) studied the issues SMEs can face in changing environment and the potential of e-commerce in new environment. Bang Nam Joen et al. (2006) investigated the determining factors of the successful adoption of e-business by small and medium enterprises in Korea, and find the important determinants of the successful adoption of e-business by SMEs in Korea are: the CEO's knowledge of information technology (IT)/e-business, relative advantages and benefits from implementing e-business, governmental support, globalization strategy and the North Korea factor[1]. Pittawa Ueasangkomsate (2015) analyzed SMEs who are exporters in Thailand to e-commerce adoption for export market and suggested Thai SMEs both exporters and non-exporters now realize great benefits of e-commerce to global market at high level [2]. Yang Jian-Zheng et al. (2013) conducted an empirical study on the factors affecting cross-border E-commerce development of SMEs, and e-marketing, electronic customs and international logistics are the impact of SMEs cross border e-commerce mode selection and most of SMEs who are weak at cross border e-commerce operation stay on third party cross-border e-commerce service platform[3].

Researches also focus on the advantages of SMEs adopting cross border e-commerce, Tao Qu (2005), Feng Chi Shen (2008), Meng Yan Wang (2014) studied the adopting of cross border e-commerce by SMEs to face the influences of a changing market environment and financial crisis. they found SMEs' adoption of cross border e-commerce can weaken the negative factors they face: more complicated trade environment, increased labor cost and rising of raw material etc.

In summary, most of the researches found that cross-border e-commerce did facilitate the development of SMEs, and there are still limitations regarding the adoption of cross-border E-commerce, this paper is going to analyze comprehensively the adoption of cross-border E-commerce in China and its influence on China's SMEs, and then discuss the reason of SMEs in China to adopt cross-border E-commerce and the challenges and limitations they are facing.

[1] Bang Nam Jeon, Kyeong Seok Han & Myung Jin Lee, Determining factors for the adoption of e-business: the case of SMEs in Korea, Applied Economics, Volume 38, 2006-issue 16 Pages 1905-1916, 21 Nov. 2006

[2] Pittawat Ueasangkomsate, Adoption E-commerce for export market of small and medium enterprises in Thailand, 11th International Strategic Management Conference 2015, Procedia-Social and Behavioral Sciences 207 (2015) 111-120

[3] Yang Jian-Zheng, Yang Na-Chuan, Yang-Li-Fan, The factors affecting cross-border E-commerce development of SMEs-an empirical study, The third Wuhan International conference on E-business-emerging operation & services management, 2013, 749-757

3. OVERVIEW OF CROSS BORDER E-COMMERCE DEVELOPMENT IN CHINA

3.1. China's cross-border e-commerce market size

In recent years, the emerging cross-border e-commerce development in China is prominent, transaction volume keeps increasing year by year, and proportion in international trade also keeps increasing, in 2008 the size of cross-border e-commerce transactions in total imports and exports is only 4.3%, this number reached to 26% in 2016. Meanwhile, cross-border e-commerce transaction volume is growing fast, the growth rate is basically maintained around 20% more or less, and in 2010 the growth rate even reached 44.6% (refer to Table 1), far exceeded the growth rate of China's GDP's (refer to Figure 2), cross-border e-commerce has become an important role in promoting China's economic growth.

In terms of total transaction volume, cross-border e-commerce in 2008 amounted to $ 109.60 billion, $ 131.64 billion in 2009 and $ 978.68 billion in 2016, 7 times higher than the volume in 2008. In 2017, the estimated transaction volume is likely to reach 7.5 trillion Yuan (about 1131.2 billion US dollars), in 2018 the volume might exceed 8 trillion Yuan, reaching 8.8 trillion Yuan (about 1327.3 billion US dollars). That shows China's cross-border e-commerce market has been making tremendous contributions for China's national economic development.

According to Aliresearch's report, this fast growing speed will continue, in 2020, China's cross border e-commerce volume is estimated to reach 12 trillion Yuan, and will be about 37.6% of China's total import and export. B2C transaction volume of cross border e-commerce will exceed 3.6 trillion Yuan with an annual growth rate of 37%, and occupy over 30% of China' s total cross border e-commerce volume[4].

China is a major import and export country, and has long been in advantageous position in international trade, if the advantages in trade can combine the rapid development of cross-border e-commerce, then this will promote China's further growth in international trade, meanwhile it will create new business opportunities for China's small and medium sized enterprises, which are normally in a position with competitive disadvantage, and bring new business growth.

Table 1 China 's cross - border e-commerce in 2008 - 2017

	Total import and export volume (in USD billions)	Cross border e-commerce volume (in USD billions)	Cross border e-commerce in total trade(%)	Cross border e-commerce growth rate(%)
2008	2563.26	109.60	4.3	N/A
2009	2207.54	131.64	5.9	20.1
2010	2972.76	190.39	6.4	44.6
2011	3641.94	271.84	7.5	42.8
2012	3866.80	365.17	9.4	34.2
2013	4160.30	492.97	11.8	35
2014	4303.00	651.17	15.1	32.1
2015	3956.90	834.89	21	28.2
2016	3685.60	978.68	26	17.2
2017E	3599.90	1131. 20	31	15.6

Sources: National Bureau of Statistics of China and published data of China e-commerce Research center

[4] Future of Trade: Cross border e-commerce connecting to the World, 2016 China's cross border e-commerce development report, Aliresearch, ACERC, 2016.9

Figure 1 Cross border e-commerce in total trade volume (2008-2017E)

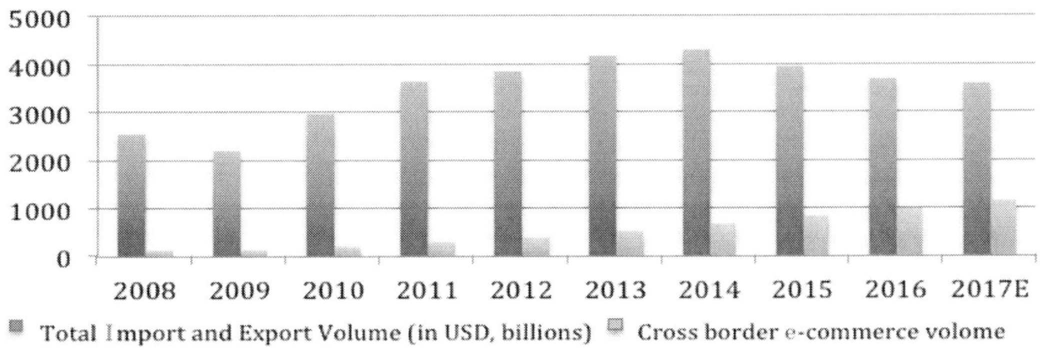

Sources: National Bureau of Statistics of China and published data of China e-commerce Research center

Figure 2 Cross border e-commerce growth rate and GDP growth rate (2008-2017E)

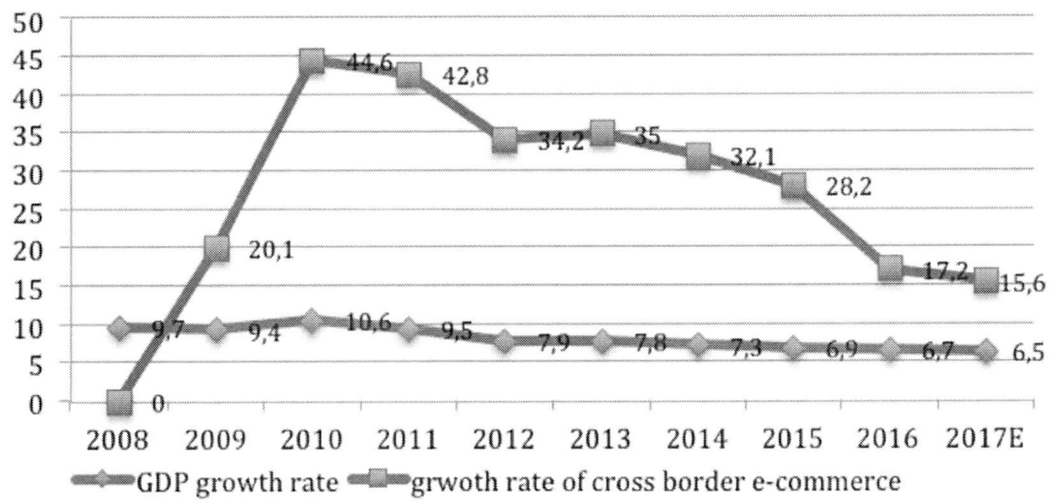

Sources: National Bureau of Statistics of China and published data of China e-commerce Research center

3.2. Structure of China's cross-border e-commerce

As cross-border e-commerce in China is relatively new, because of its large and stable trading volume, B2B was first to develop in China, and accounted for more than 90% of China's cross border e-commerce market (refer to Figure 3). However, cross-border e-commerce's fast development of, and the business success of participated enterprises, attracted more SMEs joined in the cross border e-commerce market, meanwhile with the facilitation of the development of international logistics, cross-border B2C transactions gradually become realizable, SMEs can use online direct marketing to reach and deliver the goods directly to overseas consumers overseas, therefore the scale of cross-border B2C is also increasing, in 2010 the proportion of B2C only accounted for 2.3% of the total cross-border e-commerce, but this proportion is increasing and is expected to reach 11.1% in 2017 (refer to figure 3).

Figure 3 Structure of China's cross border e-commerce mode

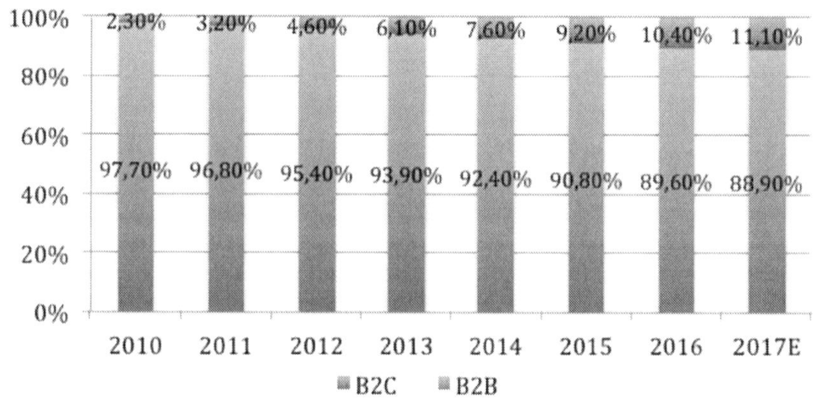

Sources: Published data of www.iresearch.cn

Although China's cross border e-commerce is experiencing a fast development, its structure is very imbalanced, proportion of cross-border e-commerce exports is much larger than that of imports, despite the downward trend of cross-border e-commerce exports, current export cross-border e-commerce still accounts for more than 80% of the total cross-border e-commerce volume, imports of cross-border e-commerce transactions are less than 20%(refer to Table 2 and Figure 4). The reason behind this imbalance structure is mainly because most of the demand of domestic buyers and consumers can be met by domestic e-commerce business and by traditional foreign trade, therefore did not rely on cross-border e-business imports. At the same time, China's large volume of SMEs, in order to expand the market, especially when experienced a sharp decrease of orders after the financial crisis, joined in cross border e-commerce to increase market share and profits, thus promoted the substantial increase of cross-border e-commerce exports.

Table 2 Ratio of China's cross border e-commerce imports and exports in total cross border e-commerce volume in 2010-2017 (unit: %)

	2010	2011	2012	2013	2014	2015	2016	2017
IMPORTS	6.5	7.6	9.8	11.8	13.3	14.6	15.6	16.2
EXPORTS	93.5	92.4	90.2	88.2	86.7	85.4	84.4	83.3

Sources: China E-commerce Research Center

Figure 4 Ratio of imports and exports in cross border e-commerce in 2010-2017E

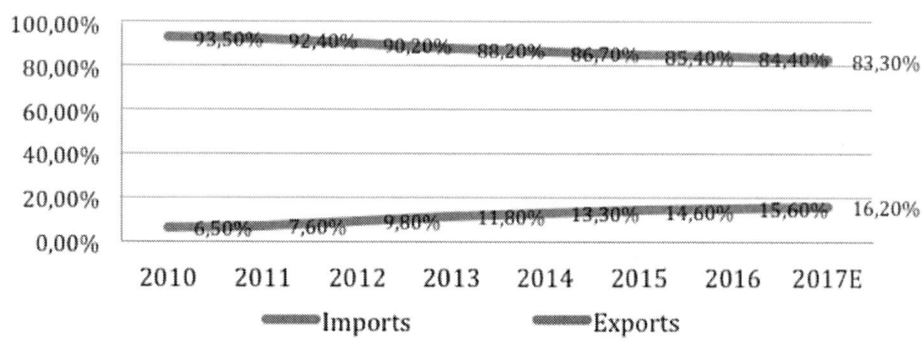

Sources: China E-commerce Research Center

4. STATUS QUO OF CHINA'S SMES IN DEVELOPING CROSS BORDER E-COMMERCES

4.1. Situation of SMEs in China to develop cross-border e-commerce

According to the National Bureau of Statistics data in 2017-2022 depth research on China's enterprise business industry, there are more than 40 million small and medium enterprises in China, accounting for 99% of the total number of enterprises, contributing 60% of China's GDP, 50% of the tax and 80 % Of urban employment. In the cross-border e-commerce market, China's small and medium enterprises are very active, up to June 2015, E-commerce platform companies are more than 5,000, personnel directly employed by e-commerce services enterprises are more than 2.55 million, employment driven by e-commerce has exceeded 18.35 million. In major business platforms in China, registered small and medium enterprises are more than 20 million, accounted about 98% of the total registered companies.

In 2013, China's e-commerce transactions totaled 10.1 trillion Yuan, of which SMEs' B2B volume occupies half of the total transaction volume, reached 51.7%, if count B2C and C2C, SMEs created a total of nearly 70% of the transaction share. In cross-border e-commerce market, SMEs also accounted for 60% of the market, and growth rate of their business reached 21.1%. Transactions in 2015 increased to 16.2 trillion Yuan, and the share in total cross border e-commerce is about 80%. In 2015, the scale of cross-border e-commerce market reached 5.3 trillion Yuan, accounting for 19.4% of the total import and export trade, cross-border e-commerce reduced the intermediate in trade, shorten the transaction cycle of trade and fasten the turnover of operating capital, and dramatically reduced the transaction cost of SMEs in trade, thus attracted considerable SMEs in cross border e-commerce and helped those SMEs increased their sales, cross border e-commerce is developing fast, and SMEs has become the main force of China's cross-border e-commerce development.

4.2. Reasons for the development of cross-border e-commerce in small and medium-sized enterprises

4.2.1. Frustration in Export

Affected by the financial crisis, the momentum of economic development of the world is no longer strong, even until now, world economy is still in recovery, and the process is slow, China's export is greatly affected, export growth started to slow down, and even showed negative growth in 2015 and 2016. Figure 5 shows that China's export volume in 2010 was 1.58 trillion US dollars, with an increase of 31.3%. However, in the last two years, China's exports have been on a downward trend, and even experienced a negative growth in 2015 and 2016, with a growth of -2.9% and -8.3%. The negative growth trend of traditional export trade growth has a huge impact on China's exports, especially on SMEs, which normally export low value-added products, and rely greatly on low manufacturing cost, but now, the cost advantage of SMEs is gradually disappearing, and their profits and overseas markets are slowly shrinking. Cross-border e-commerce has brought new opportunities for the development of SMEs. Orders in cross border e-commerce normally small

Figure 5 Total export and growth rate in 2006-2016

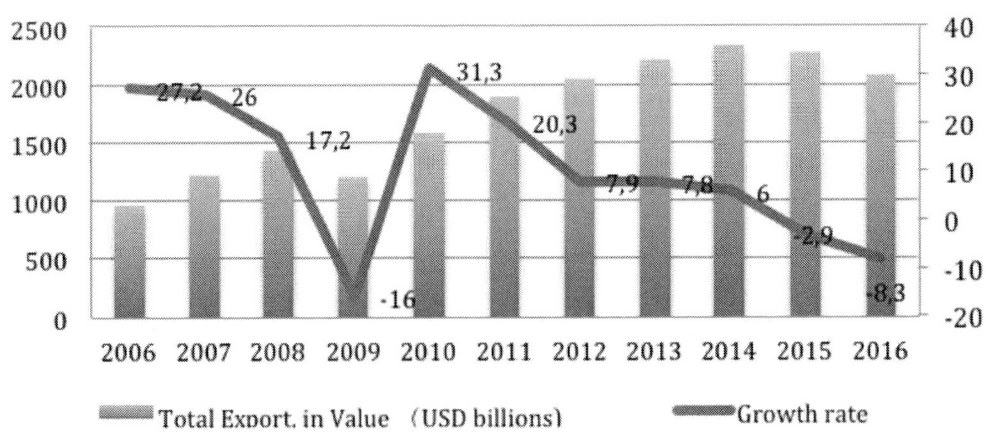

4.2.2. Cross-border e-commerce reduced the participation cost of SMEs in trade

In traditional international trade, due to the shortness of funds and scale, and limitation of cost and risk factors, there are always a lot of obstacles for SMEs when reaching overseas buyers, and that seriously affected the growth of overseas orders of SMEs. Cross border e-commerce offer a very low cost platform for SMEs to publish their products, and those platforms' huge traffic make it possible for SMEs to reach their potential customers and understand the customers' needs and immediately arrange production, thus reduce SMEs' marketing cost and inventory cost, and also make it possible for SMEs to find raw material suppliers. Meanwhile cross border e-commerce broke the constrains of time and geography, and platforms' credibility is an assurance for suppliers and buyers, products information became more actual and transparent, which made it easier to create more trade opportunities.

4.2.3. Support of preferential policies

Government's support also promoted the adoption of cross border e-commerce of SMEs. A lot of preferential policies for SMEs have been introduced to develop cross border e-commerce, including tax incentives, establishment of cross-border e-commerce pilot, etc. In July 2013, The State Council and the Ministry of Commerce officially announced "support foreign trade comprehensive service enterprises to supply financing, customs clearance and export tax rebates services" which will help solve the problem SMEs face in adopting cross border e-commerce. Ministry of Finance and the State Administration of Taxation issued "notice on cross-border e-commerce retail export tax policies"in January 2014, which confirmed the preferential tax policies for cross-border retail sales. SMEs, which apply cross border e-commerce, can enjoy a certain degree of export tax rebates in the future. In 2015, the State Council introduced a number of policies on supporting the development of cross-border e-commerce, such as "guidance on the promotion of the healthy and rapid development of cross-border e-commerce", puts forward measures in customs supervision, inspection and quarantine, import and export tax and payment settlement to support the development of cross-border e-commerce and create a good environment for supporting the development of cross-border e-commerce.

After those preferential policies, the level of customs supervision and service has been continuously improved. General Administration of Customs requires the implementation of cross-border electricity supplier supervision for 365 days throughout the year, which optimized cross-border e-commerce's customs clearance process, when the goods arrive the customs supervision sites, customs clearance will ben done within 24 hours. Inspection and quarantine procedures are simplifying. For example, Zhejiang province issued " the guidance on inspection procedure of cross-border e-commerce", simplified the inspection process, and apply different supervision methods for direct mail and bonded import. Moreover, the import and export tax system is optimizing continuously. Government will further reduce the tax for e-commerce companies, implementing the value added tax, excise tax refund or tax exemption policies.

5. ISSUES FACED BY SMES IN ADOPTING CROSS BORDER E-COMMERCE

5.1. Issues faced by SMEs

Fast development of cross border e-commerce and supports from the government did promote the growth of SMEs, but SMEs in China are still facing a lot of problems.

5.1.1. High homogenization of exporting products and vicious price competition

With the rapid development of China's cross-border e-commerce, more and more SMEs have joined the cross-border e-commerce competition, and the competition became very fierce. Some of the export products with high profit margin showed a high homogeneity, the increased number of participated SMEs worsened the completion in price and squeezed SMEs' profits, low profits make companies lack the capability to establish good customer service and build a competitive brand, therefore lack long-term development capability.

5.1.2. Relatively difficult customs clearance

Unlike the traditional import and export trade, in cross-border e-commerce, especially in B2C e-commerce, goods value are normally small, but at present, companies have to face cumbersome customs clearance, settlement and export tax rebate procedure. Lots of domestic SMEs engaging in small amount B2B business are still facing the problems in customs declaration, commodity inspection and customs clearance.

5.1.3. Long and expensive commodity logistics

The cross-border e-commerce involves many countries with different domestic situation and policies, it is difficult for SMEs to arrange the international logistics, and the process of delivery is relatively long and relatively expensive, and delivery time is not stable, especially for goods deliver to South America, Russia etc, which normally take 25-35 days. In some of the overseas markets, the logistics of goods are difficult to trace, and occasions of lost packages occurs. B2C goods normally are transported by air and the cost is relatively high, currently, the logistics cost is about 25% of China's cross border e-commerce companies' revenue.

5.2. Reasons for obstacles faced by SMEs

5.2.1. Lack of detailed and specific stipulations and policies

Due to the rapid development of cross-border e-commerce in China, China has introduced many policies and stipulations, but lack of specific details or the specification for those policies, meanwhile supportive and preferential policies lack enforcement during implementation, which is an important factor in restricting the development of cross-border e-commerce in china.

5.2.2. Lack of stable transaction payment system

In China, the domestic payment system is very advanced, but cross border online payment normally are made by using third payment party, such as PayPal, money gram and Western Union, which all have their own limitations, either contains a high transfer fee or has transactions time limits. The well domestically adopted Alipay is not world widely used, and relative supervision on payment system is not perfect at the moment, stability and security of the online payment system contains certain risks.

5.2.3. Lack of complete cross border transportation system

As an important part in trading, logistics is one of the primary factors restricting the development of cross-border e-commerce. At present, cross border logistics used by SMEs in China is mainly through Post parcel, UPS, DHL and Fedex etc., Post parcel freight is low, but deliver time is long and customer experience is relative poor; UPS, DHL and Fedex are fast but freight is high, China's existing logistics system is developing but didn't meet the high demand of the rapid developed demands of SMEs, at the high cost of transportation is a challenge for SMEs.

5.2.4. Limitation of SMEs

Because of the limitation of the sizes, SMEs in China are normally difficult to receive credit from the bank, and difficult to hire competent personnel to expand the market, reach customers in different countries, solve disputes in trading, and offer sound aftersales services, and management in those SMEs normally lack a long term development view, all those limited China's SMEs' development as well.

6. CONCLUSION

In the last ten years, cross border e-commerce's development in China is prominent, transaction volume experienced a more than ten times increase with an annual growth rate of almost 20%, the fast development of China's cross border e-commerce promoted the development of China's export and created a very good chance and platform for China's SMEs. By adopting cross border e-commerce, SMEs in China are able to reach more markets and customers and increased sales, and their development with the facilitation of cross border e-commerce is also prominent. However, due to the incomplete development of cross border e-commerce, lack of detailed policies and stipulation, loose in implementation of existing policies, lack of supervision in payment system, high transportation costs, limitation in themselves of SMEs etc., SMEs in China are still facing a lot of problems when adopting cross border e-commerce, especially when the high development trend is predicted to continue, and those problems should not be overlooked.

REFERENCES

- Mesut Savrul, Ahmet Incekara, Sefer Sener(2014), The potential of E-commerce for SMEs in a globalization environment, 10th International Strategic Management Conference, *Procedia-Social and Behavioral Sciences* 150, 35-45
- Bang Nam Jeon, Kyeong Seok Han & Myung Jin Lee(2006), Determining factors for the adoption of e-business: the case of SMEs in Korea, *Applied Economics,* Volume 38, 2006-issue 16, 1905-1916
- Pittawat Ueasangkomsate (2015), Adoption E-commerce for export market of small and medium enterprises in Thailand, 11th International Strategic Management Conference 2015, *Procedia-Social and Behavioral Sciences* 207,111-120
- Yang Jian-Zheng, Yang Na-Chuan, Yang-Li-Fan(2013), The factors affecting cross-border E-commerce development of SMEs-an empirical study, *The third Wuhan International conference on E-business-emerging operation & services management,* 749-757
- Future of Trade: Cross border e-commerce connecting to the World, *2016 China's cross border e-commerce development report,* Aliresearch, ACERC, 2016.9
- Meng Yan Wang (2014), Researches on the development of small and medium sized enterprises applying cross border E trade, *Enterprise Economy,* 81-85
- Wei Tao Chen, Jian Liu (2012), Finance restriction and export of SMEs, World Economy Studies, 43-48 Lei Zou, Ce Xu (2015), Promoting the health and fast development of cross border e-commerce, *Macro economy management,* 33-36
- Chuanxi Li (2015), Characteristics and trend of China's export cross border e-commerce development, *Special Zone Economy,* 136-138
- Institute of Economics Shanghai Academy of Social Science (2014), Study on the topic of cross border e-commerce development and its government regulation in China, *Shanghai Journal of Economics,* 3-18
- Ireserch (2016), Report of China cross border e-commerce'd development 2016.

Author Text

Li Tang, Department of International Trade and Economics, School of Economics, Tianjin University of Finance and Economics, China
Email: Tanyatte@aliyun.com

World Politics and Human Rights

What is prison for? The "professional ethos" of prison guards and education officers

Afonso de Castro Bento,

Phd Student, CRIA-IUL, Portugal

ABSTRACT

The Portuguese state has been consistently adamant in affirming the rehabilitative and humanist focus of its penal policies for the last forty years. Recently, a surge in the incarceration rate - which rose from 102 inmates per 100 000 inhabitants in 2008, to 134 in 2016[5] - coupled with a political and economic crisis, has placed this general orientation under renewed pressure. Drawing upon interviews and ethnographic fieldwork carried out in two Portuguese prisons, this paper will propose some exploratory remarks concerning the "professional ethos" which prison guards and education officers deploy in order to bring to life the system's utilitarian goal. The two prisons under analysis - Oporto and Guarda - differ in their overall characteristics and problems, making their comparison relevant for the understanding of the Portuguese penal context.

1. INTRODUCTION

The Portuguese state has been consistently adamant in affirming the rehabilitative and humanist focus of its penal policies for the last forty years. The different strains of penal law which emerged from the post-revolution democratic society in 1974 remain a point of pride among penal experts and politicians and are constantly hailed as a civilizational achievement. However, they have been constantly tried by different sets of historic circumstances, the project of a utilitarian prison being no exception. Currently, a surge in the incarceration rate - which rose from 102 inmates per 100 000 inhabitants in 2008, to 134 in 2016 - coupled with a political and economic crisis, has placed this general orientation under renewed pressure. Drawing upon interviews and ethnographic fieldwork carried out in two Portuguese prisons, this paper will propose some exploratory remarks concerning the "professional ethos" (Fassin, 2015) which prison guards and education officers deploy in order to bring to life the system's utilitarian goal. The two prisons under analysis - Oporto and Guarda[6] - differ in their overall characteristics and problems, making their comparison relevant for the understanding of the Portuguese penal context.

As Didier Fassin (2015) reminds us, the state is not an abstraction. It cannot be boiled down to a specific rationality, directives or set laws. Instead, it is the product of a dialectic relationship between the "professional ethos" and "moral subjectivities" of individuals and the prerogatives of the State and shifting "moral economies" (idem). In other words, it is created by its institutions and agents in everyday situations. Therefore, to understand prison daily life we must understand the motivations, expectations, and worldview of those who work in these institutions, and how they intersect with overarching objectives, priorities and concerns. Having that in mind, I will, firstly, present a succinct overview of the Portuguese penal system's current situation and discuss the official discourse which surrounds it. After that, I will describe the different institutional settings which will be analyzed in this paper. Finally, I will explore the "professional ethos" (Fassin, 2015) of prison guards and education officers

2. THE PORTUGUESE PENAL SYSTEM: AN OVERVIEW

There is a sort of utilitarian ethos which pervades official Portuguese discourses about prison. Politicians and penal experts will often praise the progressive and humanist nature of Portuguese penal law and highlight its short sentences, its aversion to retributive punishment, the way it assures inmates full citizenship rights and, perhaps most importantly, the way it sees social reintegration as prison's main goal. It is interesting to notice how Portuguese penal laws are frequently framed as a sort of civilizational achievement, a testimony of the nation's development and forward thinking institutions. In a sense, this attitude towards penal law seems to be a part of a broader attitude towards development on the part of political and academic elites, which base their discourses - and sometimes their policies - on a supposed need to counter a chronic underdevelopment and lack of modernization (Frois, 2013).

[5] Data from the World Prison Brief.

[6] I carried out this fieldwork when I was a research assistant in the project "Human Security: Perspectives, Subjectivities and Experiences", coordinated by Professor Catarina Frois and funded by the Gerda Henkel Foundation, to which I'm indebted.

However, it is no secret to anyone that there is a consistent lag between the promise of Portuguese penal law and the reality of Portuguese prisons. The ideal of a utilitarian prison has been continuously challenged by a changing set of circumstances since 1974. In current times, it has been affected by a trying conjecture which combines a surge in the incarceration rate with a political and economic crisis, which peaked when, in 2011, the government declared the need for a financial bailout and subsequently stepped down[7]. In between 2008 and 2016, the incarceration rate rose from 102 to 134 inmates per 100 000 inhabitants[8], causing a considerable increase of the prison population. Currently, although the system as an overall capacity for 12 600 inmates, it houses approximately 13 779 individuals. In 2007, it had room for 12416 inmates and housed 11587[9]. In short, this means an increase of 15% in less than 10 years. This, has, of course, aggravated a set of underlying issues which have affect the Portuguese penitentiary system since 1974. Many prisons do not comply with the law which mandates that there be only one inmate per cell. This means that many of the Portuguese prison establishments have two or three individuals per cell, but also, in many cases, that there even more in spaces destined for individual housing. Furthermore, the overcrowding issues in many prisons prevent the de facto triage of inmates which is deemed essential for the accomplishment of the system's goals. Custody inmates share their surroundings with convicted felons; low level offenders - convicted for sentences up to three years - share their cells with inmates who were convicted up to twenty years. Also, most of the prison establishments lack in specialized work force, namely prison guards and rehabilitation officers. In fact, during the last years the system has been plagued by countless strikes of prison guards, which demand not only the actualization of their salaries, but also the hiring of more of their colleagues. It is the case that many guards say that it is impossible for the prisons they work in to be safe places. Finally, just to give one more example, some prisons lack in occupations for inmates - be they recreational or labor-oriented - and end up inflicting a punishing lethargy upon many individuals which, again, is counterintuitive to the system's design and goals.

The new director of the Portuguese Prison Services - who entered office in 2015 - addressed these issues in a recent speech where he confronted critics who agreed that Portuguese prisons don't live up to the standards of Portuguese Law:

> "Dear friends, we can talk a lot, speak of rights and even name articles of the law. We can do that. But that that is meaningless if we really mean to address the real problem. The problem of the prison system will only be solved when, first and foremost, we address the size of our prison population (…). We have to change the attitudes of our courts or we aren't going anywhere. That's the cause of our problems. As long as we have 14 000 inmates there is no adequate solution[10]."

This excerpt reflects the position of the current administration, which claims there is a sizable proportion of inmates who wouldn't necessarily have to be in prison. Supposedly, amongst these, there are inmates which are denied access to electronic bracelets for overzealous reasons, who are imprisoned for minor offences, and who are denied early leave with no apparent justification. In short, there can be no upholding of the law if, quoting the Director of the Portuguese Prison Services, there are people who are in prison "for trivialities"[11]. The fault, therefore, must be found in the courts and in their overly punitive interpretation of the different strains of the Portuguese penal law, which, in their view, has been spiking the incarceration rate. In turn, the solution for this administration - including the current minister of justice - has been to actively promote alternatives to imprisonment, such as electronic bracelets[12].

However, this is not a new problem (neither are the solutions). Ever since 1974, there has been a contradiction between the spirit of Portuguese penal law, and the reality of criminal policy. It's not even a contradiction which resumes itself to the realm of penal law and practices. Many have noticed that there is a persistent inconsistency between the general set of laws ushered in by the revolution and the practices of the general population. For some, this is even a

---

[7] https://www.theguardian.com/business/2011/apr/06/portugal-admits-need-for-eu-financial-lifeline

[8] Data from the World Prison Brief.

[9] Data from Direção Geral dos Serviços Prisionais e da Reinserção Social. http://www.dgsp.mj.pt/

[10] This speech is from a conference organized by the Portuguese Bar Association in 2016, entitled: "Our prisons: which present and which future?" https://portal.oa.pt/comunicacao/noticias/2016/10/as-nossas-prisoes-que-presente-e-que-futuro-galeria-de-imagens/

[11] This was a said in a parliamentary hearing from April of 2016. http://www.canal.parlamento.pt/?cid=905&title=audicao-do-diretor-geral-de-reinsercao-e-servicos-prisionais

[12] http://observador.pt/2016/02/01/ministra-da-justica-defende-uso-pulseira-eletronica-reduzir-sobrelotacao-nas-prisoes/

characteristic which can describe Portuguese Society post 1974 (Sousa Santos, 1990; Ferreira, 1999; Barreto 2000; Bravo, 2000; Costa, 2003). The fact is that the democratic revolution of 1974 ushered in a pattern of increasing prison population. In itself, this is not a surprise. Litigation, instability and conflict are more common in democratic societies then in authoritarian ones (Barreto, 2000). But apart for short periods of decline, the tendency of the Portuguese prison population to increase has constituted the prevailing trend until this day. This remains true in spite of the fact that various penal codes have favored the execution of alternative sentences, such as community work, suspended sentencing or the payment of fines. In fact, although one could formulate the hypothesis that a punitive stance is becoming common in Portuguese courts, Caiado (2016) has pointed out that the system which oversees alternative sentencing has also experienced tremendous growth.

3. TWO PORTUGUESE PRISONS: OPORTO AND GUARDA

It must be made clear that this very broad panorama of the Portuguese system disguises very disparate penitentiary contexts. There are establishments which operate with high levels of overcrowding, such as Oporto prison, and others, such as Odemira which operate below capacity. There are also establishments which house a specific kind of inmate population and dispose of an adequate team of education officers which implement specialized programs - such as Carregueira, a novel prison which primarily houses inmates accused of sexual crimes -, and establishments such as Leiria, which was designed to hold only remand inmates and currently, due to considerable overcrowding, houses a very diverse population which is mainly left to itself in small-sized facilities.

The two prisons under analysis also give shape to these contrasts. Oporto prison is one of the biggest facilities in the country, being assigned the role of housing the remand male inmates of the northern region of the country. However, having an overcrowding rate of almost 200% - that is, housing 1183 inmates in facilities designed to hold only 686 individuals[13] - it operates near breaking point. It has a shortage of guards - most of the time assigning only few guards to supervise units of 200 inmates - as well as education officers - which are normally assigned over 100 inmates to monitor. Some of the units which compose the facility are described by some inmates and guards as "jungles", having a unsupervised environment. Guarda prison is very different. It is a smaller institution which was assigned the role of housing convicted male inmates of the center region. Although it houses some remand inmates, it operates at normal capacity having approximately 180 inmates (a small percentage of which are female) and almost no overcrowding[14]. Daily life is largely uneventful and a high percentage of inmates either work or study[15].

4. THE "PROFESSIONAL ETHOS" OF PRISON GUARDS AND EDUCATION OFFICERS

The second article of the code which regulates the execution of prison sentences[16] is devoted to the "finality" of those kind of measures. It states:
"The execution of sentences and measures which deprive the liberty of individuals aim to reintegrate the agent in society, preparing him to lead his life in a socially responsible manner, without committing crimes, the protection of legal assets and the defense of society."

In a way, prison guards and education officer's different functions represent the two main facets of the Portuguese penitentiary system. On the one hand, we hand we have security, order and surveillance. In the other, we have the imperative of rehabilitation and reintegration. These two set of professionals are, therefore, mandated with bringing the system imperatives to life.

As mentioned earlier, this analysis will be indebted to the perspective of Didier Fassin (2015), which argues that the State is not an abstract or unified entity whose imperatives are automatically and uncreatively carried out by its employees. It is dependent upon the individuals who work for its institutions and which, always to a certain degree, will have discretion to carry out their functions in a manner more or less consistent with their values and judgments. That is why it is so important to understand what kind of "professional ethos" prison guards and education officers deploy in order to navigate the challenges of prison life. It is a first, but inevitable step towards understanding how the

[13] Data from Direção Geral dos Serviços Prisionais e da Reinserção Social. http://www.dgsp.mj.pt/
[14] Data from Direção Geral dos Serviços Prisionais e da Reinserção Social. http://www.dgsp.mj.pt/
[15] This is true only for the majority of male inmates. The female inmates frequently complain about the lack of occupation.
[16] Law no 115/2009, 12th of October. http://www.pgdlisboa.pt/leis/lei_mostra_articulado.php?nid=1147&tabela=leis

penitentiary system and its objectives are accom plished in these particular historical circumstances. In what follows, I will explore the "professional ethos" of prison guards and education officers from Oporto and Guarda prisons in a particular light; I will be specially focused on understanding how they view the Portuguese penitentiary system's objectives and their role in accomplishing them. In order to do so, I will use a typology present by Deborah Drake (2012) in which she sketches out the main functions which are normally attributed to a penitentiary system[17]: security, rehabilitation, deterrence, general prevention and delivery of justice. I will use this typology in order to discuss and clarify the different views of both prison guards and education officers. Finally, please keep in mind that these are preliminary explorations; they are primarily clues for further research. Nevertheless, I hope they are interesting and bold enough to spark a debate.

4.1. Prison guards

Most prison guards express a narrow view of their function. "We are in charge of security and surveillance" is what most will answer when asked what their role is within the penitentiary system. It is, in other words, a legalist approach to the job. Which is why it should be surprising that most of them don't think prison does what the law sets it out to do. For the prison guards with whom I talked to in Oporto, where their function are carried out under extreme conditions, prison either neutralizes or punishes individuals. They acknowledge the spirit of the law - the successful reintegration of inmates in broader society - but they doubt the institution they serve contributes to that purpose, either because it's an impossible aim or because there aren´t enough resources to pursue that objective. Education officers - who frequently personify the reintegration objective - are viewed as "pencil pushers", people who only carry out administrative or bureaucratic tasks.

One of the heads of prison guards in Oporto prison, a woman of approximately 50 years of age which complained about the repetitiveness of her job, gave the following statement in an interview:

A: Prison does not have the desired effect, the reintegration of inmates in society. That does not exist.

Q: So what does prison accomplish? It keeps them locked up?

A: Exclusively. To keep them locked. Recidivism is so high... And its not only here, in Portugal.

Another guard, approximately 40 years old which underlined the dangerousness of the Oporto prison daily life, expressed the following view:

" Q; What do you think prison is for?

A: In the present, prison's only purpose is to scare people. Nothing else. No one leaves here better, apart from the guys who entered addicted to drugs and which are able to free themselves from it."

In Guarda, prison guards do seem to have another outlook on reintegration. This time around, although they express doubt in regard to its overall effectiveness, they do believe prison carries out, in some way, its designated function. As it was already mentioned, Guarda prison as a relatively peaceful day to day life which is barely affected by any overcrowding. A male prison guard, of over 15 years of service, claimed the following:

"Q: And does reintegration work?

No. For some it never does. Since I work here I've met some inmates - not many - three or four times. Maybe it doesn't work that efficiently. If it did, they wouldn't come back.

Q: And why do you think that there are people that leave and don't come back, and others which frequently return?

A: Most of those people [come back] because of drugs, robberies and stuff like that. Maybe they even steal mostly because of the drugs. I think. But, most of the time, they will also lack support on the outside. From the family, but not only. If they are treated here, if they are guided towards social reintegration... Maybe those who come back lack support outside."

4.2. Education Officers

Education officers from both prisons expressed similar doubts and concerns about their profession. First, they think of themselves as overburdened by bureaucratic and administrative tasks, like attending to inmate's requests and needs - such as phone calls or visitor and package arrangements - or filling out reports, such as those needed to concede paroles. Second, they have, to different degrees, reduced confidence on the possibility of prison bringing about generalized reintegration or rehabilitation. In Oporto, these officers seem to believe that their work is worthwhile if they can positively impact few inmates. A female officer of approximately 40 years old, who assigns prison's failures to a general unwillingness on the

[17] It must be said that Drake herself does not believe prison as an institution carries out any of these functions. She sees them more as an alibi which the State uses in order to legitimate prison and its real functions, which are, in her view, mainly tied to the control of troublesome populations.

part of the system to accomplish its objectives, made the following statement about a former inmate which she helped escape recidivism: "What do I say? Accomplishment [comes from such cases]. We do not believe we are going to save all these people". In Guarda prison, a male officer, of few experience and who believes that the purpose of prison is mainly to neutralize offenders and give society a sense of security, shared the same outlook on his function:

"Our mission should be to prevent prisons from being this full, to prevent people from ruining their lives. If it is possible for us to intervene in a full grown person that already has a history of bad choices and be the preponderant factor... It would take a large ego. [But] we can have an important role."

Although one female officer from Guarda prison advocated that the reintegration process is a success in most cases, she did share the conviction that, in prison, there are those inmates which have a potential to be "recuperated" and those who don't. He argued that there were some inmates which "will never change", and that they should be sent to "another place, with less perks". Other officers, like a professional from Oporto, simply argued that some inmates are not ready to change:

"We should do a correct evaluation in the beginning and check in what faze the individual is at. If he is in a changing phase, if he has it 'rock bottom' and really wants to change. We should check in what phase they enter because many are not ready for change today."

5. CONCLUSION

I would like to conclude this paper by sketching out some exploratory remarks and hypothesis to be tested in future research about the Portuguese penitentiary system. Concerning the views of prison guards, we may want to be sensitive to the way a particular environment can influence these professional's views about the institutions they serve. The Oporto guards did not believe prison fundamentally served the purpose of reintegration or rehabilitation, instead believing it either simply neutralizes or punishes individuals. The Guarda professionals, working in a less overburdened environment, were more moderate in this respect. Finally, education officers from both prisons expressed great doubt concerning the possibility of prison really accomplishing what it is set out do, expressing also, therefore, doubt concerning their own function. But perhaps most interestingly, these same officers admitted that - not necessarily because of overcrowding or resource scarcity - one should carefully choose what inmates are worthy of investing time and work. This should inspire future research concerning not only the spread of this specific outlook, but also focusing on the criteria which these professionals use in order to choose the inmates they assist. Such research could contribute to the understanding of a possible pattern of selectivity imposed upon the goal of rehabilitation and reintegration.

REFERENCES

- Barreto, António (2000). Crises da Justiça. In António Barreto (Ed.) *Justiça em Crise? Crises da Justiça*. Lisboa: Dom Quixote, 13 - 28.
- Bravo, José António Dias (2000). Justiça Penal em Portugal: Crise para além do Ruído. In António Barreto (Ed.) *Justiça em Crise? Crises da Justiça*. Lisboa: Dom Quixote, 263 - 275.
- Caiado, Nuno (2016). A Política Criminal para a Execução de Penas e Medidas. *Julgar*, 28, 213 - 237.
- Costa, Eduardo Maia (2003). A Lei Escrita e a Lei na Prática em Portugal. In António Pedro Dores (Ed.) Prisões na Europa: *Um debate que apenas começa*. Oeiras: Celta, 93 -102.
- Drake, Deborah (2012). *Prisons, Punishment and the Pursuit of Security*. London: Palgrave MacMillan.
- Fassin, Didier (Ed.) (2015). *At the Heart of the State: The Moral World of Institutions*. London: Pluto Press.
- Ferreira, Vitor Peña (1999). Sobrepopulação Prisional e Sobrepopulação em Portugal: evolução recente, situação atual e alguns factores que a explicam. *Temas Penitenciários*, 2, 3, 7 - 37.
- Santos, Boaventura de Sousa (1990). *O Estado e a Sociedade em Portugal: 1974 - 1988*. Porto: Edições Afrontamento.

Author Text

Afonso Bento is a PhD student in anthropology from CRIA-IUL, Lisbon, Portugal.
Afonso.de.castro.bento@gmail.com

Scared Off: Comey's Letters to Congress and the Impact of Uncertainty on Voter Choice

Clarissa Jaqueline Arabella Luttmann
Undergraduate Student, Leiden University College, The Netherlands

1. INTRODUCTION

Following her unanticipated defeat to republican candidate Donald J. Trump in the 2016 US-presidential election, democratic candidate Hillary Clinton claims that her loss was importantly due to former FBI director James Comey's inference in form of two letters to Congress before the election, which raised renewed suspicion about her unsafe handling of confidential information during her tenure as United States Secretary of State. Albeit the letters' effect on pre-election polls has been studied, there does not yet exist a comprehensive explanation for the interventions' supposed impact on voter behaviour.
Therefore, the following case application aims at providing a theoretical framework that illustrates how Mr. Comey's letters, by inducing candidate uncertainty, may have impacted voter behaviour and thereby the outcome of the underlying democratic decision-making-process. The paper illustrates that besides the objective alteration of voter behaviour, Comey's first document triggered an anticipatable irrational voter overreaction, and that the second notification resulted in incomplete behavioural reversal while re-activating unfavourable cognitive associations with the Clinton campaign.

2. COMEY LETTERS, COMEY EFFECT(S)

Following a testimony in July 2016 that the FBI had completed its investigation into Clinton's communication of sensitive information from a private email server during her tenure as US Secretary of State, on 28 October 2016, then-FBI-director Comey went against the institution's principle of non-interference into an election and announced in a letter to Congress that "in connection with an unrelated case, the FBI has learned of the existence of emails that appear to be pertinent to the investigation. [...] Although the FBI cannot yet assess whether or not this material may be significant, [...] I believe it is important to update your Committees about our efforts in light of my previous testimony" (Politico, 2016). As of 6 November 2016, two days before the presidential election, Comey issued a second letter stating that "based on our review, we have not changed our conclusions that we expressed in July with respect to Secretary Clinton" (Prokop, 2016).

Resulting from this, Clinton later attributed her close defeat against her opponent Trump to both Russia's "influence campaign" to undermine her candidacy and Comey's decision to notify Congress, "raising doubts" and thereby stopping her "momentum" (Chozick, 2016). In support of this supposedly 'Comey Effect,' a timing-adjusted poll aggregate based on Huffington Post data suggests an immediate vote shift of 4% from Clinton to Trump immediately after the release of the first letter [Figure 1]. In subsequent polls, the total flux remained at about 2% and could have sufficed to swing Florida, Michigan, Pennsylvania, and Wisconsin (Wang, 2016).
Source: Reprinted from the Princeton Election Consortium, by

Figure 1. The Comey Effect

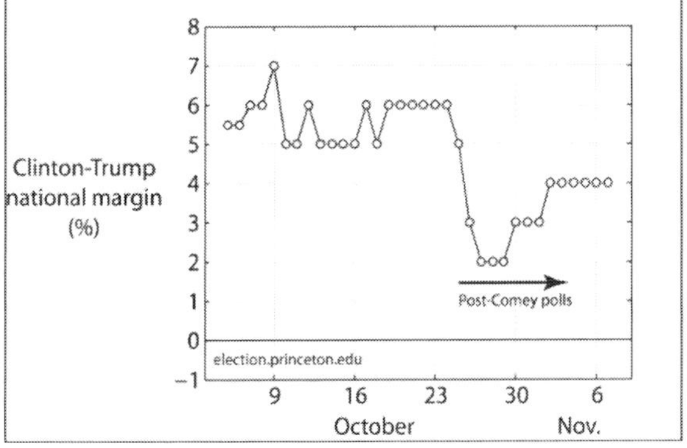

S.S.-H. Wang, 2016. Retrieved from http://election.princeton.edu/the-comey-effect-original-graph-and-post.

Clinton concluded: "it wasn't a perfect campaign, there is no such thing, but I was on the way to winning until the combination of Comey's letter on October 28, and Russian WikiLeaks raised doubts in the minds of people who were inclined to vote for me, but got scared off" (Foran, 2017).

3. ESTABLISHING THE MODEL

In order to adequately depict the letters' possible impact on the vote distribution and final election outcome, it is fundamental to first establish a model of a democratic mass election over two major candidates and in a state prior to the publication of the initial letter on 28 October 2016.
The underlying case can be illustrated by a simplifying spatial model which assumes a voter popula

tion ($N=\{1,2,\ldots,n\}$) that holds complete, transitive and single-peaked preferences over policies relating to a unidimensional spectrum of political ideology ($P=\{p,q,r,\ldots\}$) (Shepsle, 2010, p. 18). Voters are presented with alternative outcomes attached to two competing candidates' communicated positions along that same spectrum ($X=\{A\ wins, B\ wins\}$) and ($P_C=\{P_A,P_B\}$; $P_C=X$), and choose from a set of instruments with which to achieve their most-preferred, feasible outcome ($I=\{vote\ A, vote\ B\}$). Individual votes are aggregated by an electoral rule (R=plurality rule) with the implication that a vote for either candidate increases their respective likelihoods of victory.

Moreover, candidate positions are completely and identically known to the voter population, who has full certainty that once elected into office, each candidate will not deviate from their campaign-communicated ideology (Hinich and Ordeshook, 1970, p. 431). As a result, the rational individual votes so as to maximize his or her expected utility by minimizing the quadratic loss function of the spatial discrepancy between their most-preferred point and the candidate's stated position, $(P_n-P_C)^2$. That is, each voter chooses the candidate of greatest ideological proximity (Shepsle, 2010, p. 92).

Figure 2. Unidimensional Median-Voter Model

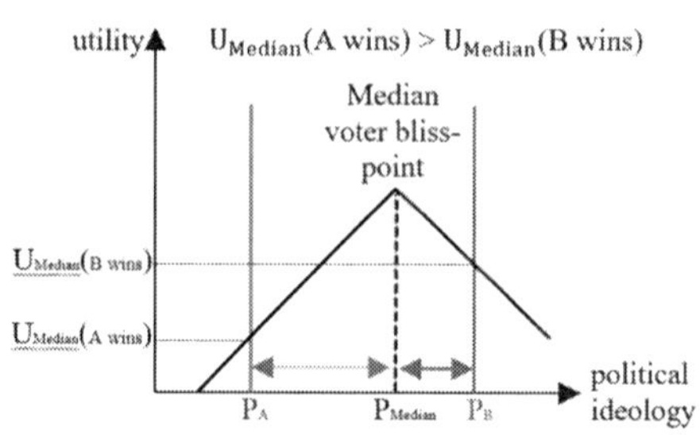

Finally, in application of the Median-Voter-Theorem, the median voter's bliss point, P_Median, has an empty win-set and therefore determines the election outcome (p. 96) [Figure 2].

4. FIRST LETTER

4.1 A Rational Reaction to Uncertainty

The first letter's impact can be analysed by introducing uncertainty about candidate Clinton's eventual ideological position if she was elected president, into the voter utility function. If the document contains no information that voters perceive relevant to the election, voter decision-making would remain unaltered. To the contrary, changes in the applicable informational environment translate into changes in voter behaviour (McCubbins and Thies, 1996, pp. 4-5). This is the case because learning, such as through the first letter to Congress, continuously adjusts voter beliefs, the "probability numbers assigned to the potential outcomes of an alternative action" (Shepsle, 2010, pp. 31-33).

Incoming information about the FBI's renewed investigation into Clinton's email server raises doubt in her overall trustworthiness and/or competence, attributes that are projected to her propensity to realize campaign promises if elected president. In the underlying spatial representation, the presence of candidate uncertainty causes the distribution of voter expectations about Clinton's actual ideological position to increase in variance (Munger and Munger, 2015, p. 167) [Figure 3].

Figure 3. Introduction of Candidate Uncertainty

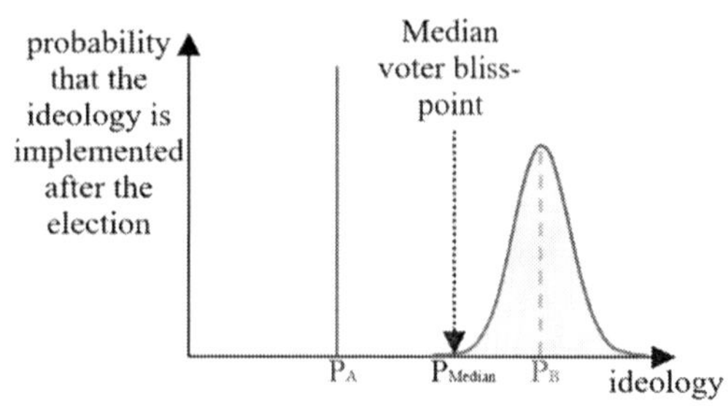

Furthermore, the expected utility maximizing behaviour becomes that in favour of the vote that maximizes the sum of the utilities associated with the outcomes that may result from the action, multiplied by the probability of these outcomes occurring (Shepsle, 2010, pp. 16, 33). Accordingly, the median voter intends to minimize the square of the difference between his most-preferred point and the candidates' perceived mean ideological position, plus the squared variance attributed to that candidate's proclaimed ideology which is assumed identical for all n, $(P_{Median}-P_C)^2+(\sigma_{Clinton})_2$. (Munger and Munger, 2015, p. 169; Morgenstern and Zechmeister, 2001, p. 98).

Because the variance of candidate Clinton's anticipated behaviour increases, voters' overall expected utility of supporting her declines, whilst the expected utility from voting for Trump remains unaltered. In cases where a voter prefers Clinton over Trump afore, $(P_n-P_{Clinton})^2<(P_n-P_{Trump})^2$, and the new element of candidate uncertainty outweighs that voter's original preferential disparity between the two candidates, $[(P_n - P_{Trump})^2 - (P_n - P_{Clinton})^2] < (\sigma_{Clinton})^2$, the vote shifts from Clinton to Trump. Consequently, only if Clinton previously enjoyed majority support, that is, if $(P_{Median} - P_{Clinton})^2 < (P_{Median} - P_{Trump})^2$ and $[(P_{Median} - P_{Trump})^2 - (P_{Median} - P_{Clinton})^2] < (\sigma_{Clinton})^2$, then, Comey's first letter would have altered the expected preliminary election outcome.

4.2. Overreaction

Moreover, the first letter may have disproportionately impacted the campaign and in ways that exceed the objective updating of candidate assessments that are subjected to rational ignorance and bounded rationality that were foreseeable by the letter's originator. Most importantly, rational ignorance applies to mass elections where n exceeds (hundred) millions of prospective voters and the objective impact of any one individual's choice is incremental (Shepsle, 2010, p. 298). In these instances, the (expected) cost of collecting, consuming and processing information about the candidates that would help make a better decision is high and often outweighs the (expected) benefit thereof (Jones, 1999, p. 309; Lodge and Taber, 2013, p. 4). In consequence, a voter will consciously ignore part of the assessable information and apply heuristics or cognitive shortcuts to this limited amount of information. To illustrate, Somin (2013) writes that at least 35% of voters are "know-nothings" (pp. 17-37). Likewise, Kahneman and Tversky (1979) find that "people rely on a limited number of heuristic principles which reduce the complex tasks of assessing probabilities and predicting values to simpler judgemental operations [...] but sometimes they lead to severe and systematic errors" (Kahneman, 1974, p. 1124).

In the underlying case, bounded rationality importantly relates to source-overestimation and false causality. Lupia, McCubbins and Popkin (2000) find that both the source-credibility and content-resonation of campaign informative sources play an essential role in (in)forming voter opinion (p. 110). With the information originating from the director of the FBI, to the general public, its subject matter likely appears of especial urgency or severity, specifically when taking the institution's beforementioned tradition of electoral no-interference into account. Furthermore, with restricted knowledge about the investigation and its source-specific association with serious criminal activity, individuals subconsciously construct a narrative that is based on associative coherence rather that objective and factual information alone (Kahneman, 2011, pp. 75, 87-88). This results in an overestimation of $(\sigma_{Clinton})^2$.

In like manner and according to Kahneman (2011), "people tend to assess the relative importance [and truthfulness] of issues by the ease with which they are retrieved from memory – and this is largely determined by the extent of coverage in the media" (pp. 8-9). Hence, a combination of both false causality and cognitive strain that mutually re-enforce each other, stipulates an "availability cascade" in which "danger is exceedingly exaggerated as the media compete for attention-grabbing headlines" (p. 143). Lastly, the public estimation of (candidate) uncertainty is subjected to risk-aversion as well as overestimation bias, the tendency of people to overestimate the relative importance of losses compared to gains of equal value, and to overestimate the probability of improbable outcomes and underweight outcomes of higher certainty (p. 32). These factors further increase Clinton's perceived ideological variance and thereby exaggerate voter reaction.

5. SECOND LETTER

5.1. Incomplete Updating

Notwithstanding, since the second letter announces that no additional evidence against Clinton has been found, and thereby negates the informational substance of the first, in a fully rational decision-making process a complete reversal of the initial change in voter behaviour should be expected. To the contrary, Clinton claims that the second letter was not helpful but even more detrimental to her campaign then the first (Chozick, 2016). One possible explanation why the 2% vote swing may have not fully reversed is that people are 'incomplete Bayesians' and "find it very difficult to escape the pull of their prior attitudes and beliefs, which guide the processing of new information in predictable and sometimes insidious ways" (Lodge and Taber, 2013, p. 169).

Impacted by confirmation and disconfirmation bias, voters tend to selectively seek and acknowledge evidence in support of existing narratives and to dismiss disconfirming evidence (Brennan, 2016, p. 43; Jones, 1999, p. 307; Kahneman, 2003, pp. 21-22). Moreover, given that the supplement had been issued nine days following the initial release and only two days prior to the election, leaving disproportionally more time for the formation than for the corrective adjustment of voter opinion, this bias's impact is especially strong. Also, that news media tend to address consumer risk aversion by focussing their coverage on negative or shocking events further limits the process of updating voter belief systems. This problem is stressed by Brennan (2016) who criticizes that people "might find a mistaken theory more interesting or fascinating than a true one" (Brennan, 2016, p. 36).

5.2. Negative Priming

Despite, exposure to the informational update is characterized by a linguistic focus on the key words such as "Clinton," "e-mails" and "FBI investigation." These key words relate to past concepts that were affectively charged in connection with candidate uncertainty and subconsciously stored in long-term memory (Kahneman, 2003, p. 52; Lodge and Taber, 2013, p. 74). Immediately upon repeated exposure, these primes are activated and bring unfavourable associations back into voter short-term memory rather than re-adjusting them [Figure 4].

Source: Reprinted from The Rationalizing Voter (p. 19), by M. Lodge and C. Taber, 2013, Cambridge: Cambridge University Press.

Figure 2. Unidimensional Median-Voter Model

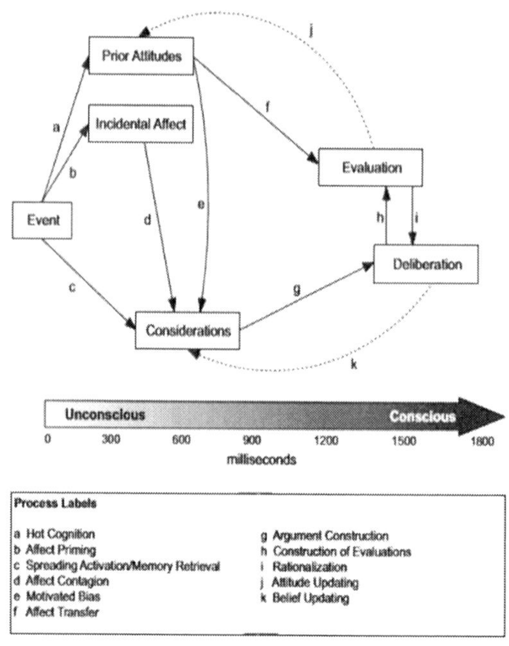

6. CONCLUSION

To conclude, this case application develops a theoretical conceptualization of the supposed 'Comey Effect' by the introduction of candidate uncertainty into the rational voter model of a democratic decision-making process. Importantly, it neither provides sufficient support of the thesis that Mr. Comey's interference factually altered the outcome of the 2016 US presidential election, nor does it evaluate the legitimacy or intention of his involvement. However, the paper outlines the theoretical conditions under which his letters to Congress sufficiently impacted voter behaviour to affect the election outcome. As a result, the power and responsibility that modern gatekeepers to information and non-information possess, especially in shaping social choice, become highly evident and underline the need to better understand and protect against such subtle vulnerabilities of the democratic process.

REFERENCES

- Brennan, J. (2016). *Against Democracy.* Princeton: Princeton University Press.
- Chozick, A. (13 November 2016). Hillary Clinton Blames F.B.I. Director for Election Loss. *The New York Times.* Retrieved 21 July 2017, from https://www.nytimes.com/2016/11/13/us/ politics/ hillary-clinton-james-comey.html.
- Davis, O., Hinich, M.J., and Ordeshook, P.C. (1970). An expository development of a mathematical model of the electoral process. *American Political Science Review, 64*(2), 426-448.
- Foran, C. (2 May 2017). Why Hillary Clinton Thinks She Lost the Election. *The Atlantic.* Retrieved 21 July 2016, from https://www.theatlantic.com/politics/archive/2017/05/hillaryclinton-trump-fbi-russian-hacking/525183/.
- Hinich, M., and Munger, M. (1992). A Spatial Theory of Ideology. *Journal of Theoretical Politics, 4*(1), 5-30.
- Jones, B. (1999). BOUNDED RATIONALITY. *Annual Review of Political Science, 2*(1), 297-321.
- Kahneman, D. (2011). *Thinking, fast and slow.* New York: Farrar, Straus and Giroux.
- Kahneman, D., and Tversky, A. (1979). Prospect theory: An analysis of decision under risk. *Econometrica: Journal of the Econometric Society, an International Society for the Advancement of Economic Theory in Its Relation to Statistics and Mathematics,47*(2), 263-291.
- Lodge, M., and Taber, C.S. (July 2013). *The Rationalizing Voter.* New York: Cambridge University Press.
- Lupia, A., McCubbins, M., and Popkin, S. (2000). *Elements of Reason: Cognition, choice, and the bounds of rationality* (Cambridge studies in political psychology and public opinion).
- McCubbins, M.D. and Thies, M.F. (1996). Rationality and the Foundations of Positive Political Theory. *Leviathan 19,* 7-32.
- Morgenstern, S., and Zechmeister, E. (2001). Better the Devil You Know Than the Saint You Don't? Risk Propensity and Vote Choice in Mexico. *The Journal of Politics, 63*(1), 93-119.
- Munger, M.C. and Munger, K.M. (2015). *Choosing in Groups: Analytical Politics Revisited.* New York: Cambridge University Press.
- Politico Staff (28 October 2016). Full Text: FBI letter announcing new Clinton review. *Politico Staff.* Retrieved 21 July 2017. http://www.politico.com/story/2016/10/full-text-fbi-letter-announcing-new-clinton-review-230463.
- Potthoff, R., and Munger, M. (2005). Voter Uncertainty Can Produce Preferences with More Than One Peak, but Not Preference Cycles: A Clue to the Fate of Ross Perot? Journal of Politics, 67(2), 429-453.
- Prokop, A. (6 November 2016). FBI director: new emails don't change our conclusion that Clinton shouldn't be charged. *Vox.* Retrieved 21 July.https://www.vox.com/2016/11/6/13542578/fbi-letter-hillary-clinton-email.
- Shepsle, K. (2010). *Analyzing Politics: Rationality, Behaviour, and Institutions* (2nd ed., The new institutionalism in American politics series). New York, NY: Norton.
- Somin, I. (2013). *Democracy and Political Ignorance. Why Smaller Government Is Smarter.* Palo Alto: Stanford University Press.
- Wang, S.S.-H. (2016). Princeton Election Consortium. *The Comey effect (original graph and post).* Retrieved July 21, 2017, from http://election.princeton.edu/the-comey-effect-original-graph-and-post.

Democracy is now regarded as a "universal good" but does it have universal understanding and definition?

Anastasia Kononova,
Master of International relations, Saint Petersburg State University, Russia

Every human community, whether family, association or nation works out the rules which govern its life. Moreover, these rules should be complied with all participants and everyone should support these rules and accede to requirements, otherwise a family life will be full of scandals and may end up with divorce, and in a state one could observe conspiracies, tyrants, revolution. Throughout the centuries-old history of humanity various forms of state rule and various political regimes were used to organize people's life. Some of them failed, some were successful enough, but the ideal one is not found so far. However nowadays democratic way of ruling the state is considered as perhaps not ideal, but very close to it. And many countries all over the world are building the democratic society and implement universal democratic values.

Democracy is now one of the most often-used concept. It is used both in political and scientific discourses. And it is widely assumed that for our society and for political system democracy is an absolute blessing. In the international relations theory was even developed a conception of liberal piece, which is also called theory of democratic piece. The core point of this theory narrows down to the assumption that states with democratic way of ruling don't declare war each other and don't fight with each other. The idea that global democracy would provide a solid foundation for global peace was stated in 1918 by Woodrow Wilson as part of his vision for a new world order and his Fourteen points also elaborated the ideas of liberalism. After collapse of the Soviet Union which also meant the collapse of communist non-democratic block, Francis Fukuyama in the year 1992 added a famous argument that humanity had reached "the end of history." And there will be no more conflicts and wars. According to the information given in Oxford Bibliographies, "The democratic peace proposition is connected to many other propositions linking domestic politics and international relations, including that democracies are more likely to cooperate with each other, that democracies are more likely to win the wars they fight, that escalating military casualties degrade public support for war, that leaders initiate conflict to secure their domestic hold on power (the diversionary hypothesis), that democracies fight shorter wars, that different kinds of democracies experience different kinds of conflict behavior, that different kinds of authoritarian systems experience different kinds of conflict behavior, and others"[1].

But what is democracy? How do we understand it? It is well known that this word and this idea came to us from the Ancient Greece. And the word democracy consists of two parts: «demos» – people и «kratos» – rule. This could be understood as «rule by the (simple) people». Famous Greek philosopher Aristotle said that democracy arises from the belief that people, equally free, must be equal in all respects. For Aristotle, democracy is not the best form of government. But since tyranny and oligarchy could be considered as even worse forms, democracy has a right to exist. Democracy – it is when the power is concentrated in the hands of the destitute and they neglect the interests of the riches. Another famous ancient Greek person, one of the "founding fathers" of Athens democracy, Pericles, affirmed: «... Our form of does not imitate the laws of neighboring states. On the contrary, we are rather a model to others. Our form of government is called a democracy because its administration is in the hands, not of a few, but of the whole people. In the settling of private disputes, everyone is equal before the law"[2]. Here, in the statement that "everyone is equal before the law" we can notice the dawning of the main thesis of the modern democracy.

Nevertheless many changes have taken place since that times, and to the XXI century many new concepts of democracy and understanding of what is democracy have been appeared, starting from the ideas of English Civil War to modern vision the model

[1] Reiter D. Democratic Peace Theory // http://www.oxfordbibliographies.com/view/document/obo-9780199756223/obo-9780199756223-0014.xml (Accessed 25.09.2016)

[2] The Golden Age of Athenian Democracy under Pericles // http://www.rjgeib.com/thoughts/athens/athens.html (Accessed 16.09.2016)

of the US as a pattern of democracy. And nowadays many modern authors make various researches concerning problem of democracy. Some of them concentrate on theoretical approaches to democracy. Such authors as L. Ali Khan[3], David Held[4] rethink and develop western concepts of democracy, others for instance, Pierre Rosanvallon[5], Jacques Ranciere[6] theorize about the possibilities and problems of democracy in terms of current evolvement.
As the predicted by Fukuyama "End of history" did not happen and the international society still face a number of military conflicts many scientists address oneself to the topic of the importance and role of democracy as a peacekeeping element. Among them one could name Bruce Russett[7], Michael W. Doyle[8], Spencer R. Weart[9].

Others take into consideration cultural differences and analyze the development of democracy regarding fundamental regional contrasts. One could name such authors as Larry Diamond, Marc F. Plattner, Yun-Han Chu, who debate the desirability of democracy in East Asia[10]. Dirk Kotze, Francis Nwonwu, who examine the prospects of continued democratic stability through the experience of 5 African countries[11]. The problem of democracy formation in Russia and related issue of capability of coexistence democracy and corruption are widely studied by Russian authors such as Kochetkov V.P and others[12]. So in this essay I would like to contemplate the existing approaches to the democracy, to understand if there is a universal knowing of this concept or not. Speaking about democracy it is useful to point out what it is definitely stays away from it. Democracy could be "defined opposite so other types of government:

- Monarchy (government by a single ruler (king/queen, emperor);
- Aristocracy (government by noblemen (hereditary);
- Oligarchy (government by few persons);
- Theocracy ("Government by God" (in reality this means government by religious leaders);
- Dictatorship (government by people, that have seized power by force (often: military dictatorship)"[13].

One of the platforms for discussions about democracy in both theoretical and practical spheres is a Journal of Democracy, which was established in 1990. It covers all various aspects of this notion. As the Journal yourself sees its mission, it is included in "explore in depth every aspect of the establishment, consolidation, and maintenance of democracy, including political institutions, parties and elections, civil society, ethnic conflict, economic reform, public opinion, the role of the media, and constitutionalism. It covers not only practical political matters but also questions of democratic theory and culture. While it maintains the highest scholarly standards, it is written and edited for the general reader"[14].

Referring to this journal we can find modern definitions of democracy which are offered by different authors. For example, Amartya Kumar Sen writes that "we must not identify democracy with majority rule. Democracy has complex demands, which certainly include voting and respect for election results, but it also requires the protection of liberties and freedoms, respect for legal entitlements, and the guaranteeing of free discussion and uncensored distribution of

---

[3] Khan Ali L. A Theory of Universal Democracy: Beyond the End of History // Martinus Nijhoff Publishers, 2003

[4] Held D. Political Theory and the Modern State.Essays on State, Power, and Democracy // Stanford University Press, 1989

[5] Rosanvallon P. Counter-Democracy:Politics in an age of distrust// Cambridge University Press, 2008.

[6] Ranciere J. Hatred democracy // Verso, 2006.

[7] Russet B. Grasping the Democratic Peace: Principles for a Post–Cold War World // Princeton University Press, 1993. Russett,B. and John R. Oneal. Triangulating Peace: Democracy, Interdependence, and International Organizations // New York: Norton, 2001.

[8] Doyle M.W. Liberal Peace: Selected Essays // New York: Routledge, 2011.

[9] Weart S. R. Never at War: Why Democracies Will Never Fight One Another // Yale University Press, 1998.

[10] Diamond L., Plattner Marc F., Chu Yun-Han. Democracy in East Asia: A New Century// The Johns Hopkins University Press, 1998.

[11] Kotze D., Nwonwu F. African Political Elites. The search for democracy and good governance // Africa Institute of South Africa, 2008.

[12] Kochetkov V.P. Philosophy Of The Russian Constitution: Russian Values And Democracy // Sententia. European journal of humanities and social sciences, №2, 2013

[13] A Short Definition of Democracy // Democracy building // http://www.democracy-building.info/definition-democracy.html (Accessed 16.09.2016)

[14] About the Journal of Democracy // Journal of democracy official website// http://www.journalofdemocracy.org/about (Accessed 16.09.2016)

news and fair comment. Even elections can be deeply defective if they occur without the different sides getting an adequate opportunity to present their respective cases, or without the electorate enjoying the freedom to obtain news and to consider the views of the competing protagonists"[15]. This author assumes that democracy comprises many aspects and each of them is critically important. Another researcher, French professor Pourhiet claims that "it is difficult to imagine that in the legal definition of democracy can be included other elements in addition to the traditional criterion of the collective sovereignty expressed in the decisions made by the majority of citizens through the implementation of the right to vote by a referendum or through elections of representatives"[16]. From her point of view it is possible to emphasize one core item which expresses quite clear the idea of democracy. And she does not give an eye to the civil liberties and human rights.

Juan J. Linz and Alfred Stepan in their book "Problems of Democratic Transition and Consolidation: Southern Europe, South America, and Post-Communist Europe" focus on the transaction of totalitarian regimes towards democracy. "A democratic transition is complete when sufficient agreement has been reached about political procedures to produce an elected government, when a government comes to power that is the direct result of a free and popular vote, when this government de facto has the authority to generate new policies, and when the executive, legislative and judicial power generated by the new democracy does not have to share power with other bodies de jure"[17]. A completed transaction means "pure" democracy. As one could notice these authors also concentrate more on way of organizing the government, not on the way of organizing the whole society.

As we got sight of, there are different opinions in academic environment of what must be included in the definition of democracy. And now we'll turn our attention to political environment and examine what do politicians comprehend in this notion.

Israel Prime-Minister Benjamin Netanyahu in one of his speeches mentioned that: "Israel is a Western liberal democracy and as such its public space is open and safe for all, men and women…there is no place in [Israel] for discrimination"[18] Here we can notice a reference to human rights and equality – characteristics of the democratic society.

Former Mayor of London and now minister of Foreign Affairs in Great Britain Boris Johnson said that: "democratic self-government – parliamentary democracy – is what the modern British nation is founded on. The most precious thing this country has given the world is the idea of parliamentary democracy. Yesterday, I believe the British people have spoken up for democracy in Britain and across Europe"[19]. His vision is that democracy is an absolute blessing inasmuch as this speech was dedicated to the Brexit, for him one of the key points of democracy is people's right to express their will.

The president of Russian Federation Vladimir Putin, answering during the interview to the question about democracy in Russia, said: "Democracy means the power of the people and the opportunity to influence the government. Russia has enough experience with a one-party system, we are not going back. We will continue to develop and improve our democracy"[20]. And he also mentioned, that there are no two identical democracies, it differs in each country and "there is no universal model of democracy". In this approach to defining democracy one can notice the ancient idea of "power of people" and if we pay attention to the rhetoric, it is interesting that he used the word "opportunity", not "ability" speaking about people's influence the government. Maybe that is one of the reasons, that for instance in Britain people really influence on politics while in Russia they just have "opportunity" and the establishment of democracy is still in process.

---

[15] Amartya Kumar Sen. Democracy as a Universal Value // Journal of Democracy, July 1999 // http://www.journalofdemocracy.org/articles/tables-contents/July-1999

[16] Pourhiet A.-M. Le. Définir la démocratie // Rev. française de droit constitutionnel. – P., 2011. – N 87.

[17] Linz J. J. and Stepan A. Problems of Democratic Transition and Consolidation: Southern Europe, South America, and Post-Communist Europe // JHU Press, 1996. – P.3

[18] Netanyahu: Israel Is a Democracy and Won't Tolerate Discrimination //Haaretz, 25.12.2011 // http://www.haaretz.com/israel-news/netanyahu-israel-is-a-democracy-and-won-t-tolerate-discrimination-1.403473 (Accessed 16.09.2016)

[19] Moore C. The European elite forgot that democracy is the one thing Britain holds most dear // The Telegraph, 24.06.2016 // http://www.telegraph.co.uk/news/2016/06/24/the-european-elite-forgot-that-democracy-is-the-one-thing-britai/ (Accessed 16.09.2016)

[20] 2.Teil Des Grossen Bild-Interviews Mit Putin „Wir wollen keine Supermacht sein"//Bild, 12.01.2016 // http://www.bild.de/politik/ausland/wladimir-putin/interview-mit-dem-russischen-praesidenten-44096422.bild.html (Accessed 16.09.2016)

And I would like to end up with the opinion of Abraham Lincoln, who of course does not belong to the modern world and in his times democracy in its current vision was just forming but that's why his words are more significant nowadays and help better understand such difficult concept as democracy. He stated, that "elections belong to the people. It's their decision. If they decide to turn their back on the fire and burn their behinds, then they will just have to sit on their blisters" [21]. Again, this quotation stresses the most important part of democracy: rule of people. In this part all the above mentioned persons agreed.

Despite the existence of different approaches to the definition of democracy we can still emphasize the basic principles of democracy, which are shared both by scientists and politicians:

- people is acknowledged as a source of power; sovereignty of the people is expressed in the fact that people elects its representatives, and can occasionally replace them, has the right to participate in the development and adoption of laws using people's initiatives and referendums;
- authorities are elected (persons exercising supreme authority are elected for a specific, limited period of time); it enables the public to monitor the government;
- competitive multi-party system; democratic regime creates the appropriate legal conditions for the functioning of political parties;
- legal political opposition, which is able to firmly criticize the government without fear of being disbanded and suppressed;
- real freedom of speech, the ability to express their views on any issue as a protection of the society from the stagnation and decay
- the equal rights for citizens to participate in ruling the country (at least equality in the right to vote);
- the minority shall abide by the majority's decision, granting safeguards for the rights of the minority
- the principle of separation of powers into legislative, executive and judicial branches;
- priority of human rights over the rights of states;
- tendency to the creation of civil society and the rule of law, and the government itself contributes to the development of forms of public selector, realization of the rights and freedoms of members of the public, compliance with laws.

In today's world system the United Nations Organisation (UN) plays an important role as an instrument of justice, peace and international cooperation, and is also based on "principle of the sovereign equality of all its Members"[22], which is one of the main pillars of democracy. The opening words of the Charter are the following: "We the Peoples". They reflect the fundamental principle of democracy, "that the will of the people is the source of legitimacy of sovereign states and therefore of the United Nations as a whole"[23]. One can say that the UN is an attempt to create a global democratic ruling. How the unfair way of forming the Security Council with 5 permanent members correlates with the above mentioned principle of equality of all the members of the UN - that's a different matter. For us it is more interesting to have a look at how the idea and definition of democracy is presented in terms of this organization. According to the information available on the official website, "democracy, and democratic governance in particular, means that people's human rights and fundamental freedoms are respected, promoted and fulfilled, allowing them to live with dignity"[24]. The principle of democratic governance embodied in the documents and statements of the UN is further replicated in national legislation. Indeed, Art. 1 of French Constitution of 1958 indicates that France is a "democratic and social republic". According to the German Basic Law 1949, Article 20, the country is "democratic and social federal state". Art. 1 of the Spanish Constitution 1978 claims that Spain is a "social and democratic state". The same about Italian Constitution 1947: Art. 1 says that Italy is a "democratic republic founded on labour"[25]. And in each country the sovereignty belongs to the people. That's what is proclaimed in basic laws, in black and white. But how it works in real life?

Nowadays some research centers work up graphics, lists and rankings about the level of democracy, its

[21] Cit.ex.Quotes on Elections. Forbes, 29.04.2014 // http://forbesindia.com/article/thoughts/quotes-on-elections/37494/1 (Accessed 16.09.2016)

[22] UN Charter, article 2, 1945 // http://www.un.org/en/sections/un-charter/chapter-i/index.html (Accessed 17.09.2016)

[23] Democracy and the United Nations //Global Issues // http://www.un.org/en/globalissues/democracy/democracy_and_un.shtml (Accessed 17.09.2016)

[24] Democracy and the United Nations //Global Issues // http://www.un.org/en/globalissues/democracy/democracy_and_un.shtml (Accessed 17.09.2016)

[25] All constitutions available online in English at .legislationline.org // http://www.legislationline.org/documents/section/constitutions

quality among countries. Countries are divided in several groups depending of how high or low their level of developing democracy is. For example, according to the Global Democracy Ranking there are 5 stages of the quality of democracy in the world from very low to very high, and the number of those states which have 100-60 scores (high level) is 51[26]. Taking into account that overall in this research were presented 113 countries, it is a good rate.

And they used the following formula to rank the democracies: Quality of Democracy = (freedom & other characteristics of the political system) & (performance of the non-political dimensions). The latter includes: (1) gender (socio-economic and educational gender equality); (2) economy (economic system); (3) knowledge (knowledge-based information society, research and education); (4) health (health status and health system); (5) environment (environmental sustainability)[27]. It is important to notice that political dimension is more significant during preparation process as it weights 50% of all indicators.

Another group of researchers, from the Economist Intelligence Unit, which also made such type of rating based their analysis "on five categories: electoral process and pluralism; civil liberties; the functioning of government; political participation; and political culture. Countries are placed within one of four types of regimes: full democracies; flawed democracies; hybrid regimes; and authoritarian regimes"[28]. They also calculated scores and estimated that 79 countries of 167 can be called democracies, and where only 26 are "full democracies". Comparing to the aforecited rating this one takes into consideration only political aspects. The first one (Global Democracy Rating) even claiming to pay more attention to political dimensions, is more specified on civil rights and freedoms. The Economist Intelligence Unit's Index also stresses the significance of civil liberties. But also states that "they [free and fair elections and civil liberties] are unlikely to be sufficient for a full and consolidated democracy if unaccompanied by transparent and at least minimally efficient government, sufficient political participation and a supportive democratic political culture. It is not easy to build a sturdy democracy. Even in long-established ones, if not nurtured and protected, democracy can corrode"[29]. So as we could have noticed before, speaking about democracy first of all we think about political way of governing, keeping in mind universal human rights and freedoms. And if the democracy was once established it does not mean that the political system is not going to change. Here is hidden one of the main problems in defining democracy – it is changeable phenomenon. And if for example we define this notion in one way and after that it will change - would it mean that it is still a democracy, or we should think about another word to describe it. And where are the strict borders to differ "full democracy" from "low democracy" – they both are democracies, but definitely the political system and human rights in Moldova are far cry from the same in Norway. All these aspects bring more mess in an arduous process of adopting universal definition of democracy.

Summing it up I would like to cite a famous politician, former British prime-minister Winston Churchill who once said "There is no such thing as the "perfect form of government" on earth, but any other form of government produces even less desirable results than democracy. Until today, no other form of government has been invented that could regulate public affairs better than democracy"[30]. And even if we do not have a universal definition of what is democracy we still have a universal understanding that democratic principles and pillars, such as: sovereignty of the people, government based upon consent of the governed, majority rule, minority rights, guarantee of basic human rights, free and fair elections, equality before the law, due process of law, constitutional limits on government, social, economic, and political pluralism, values of tolerance, pragmatism, cooperation, and compromise, - are essential parts of the right way of organizing people's life and impartial way of ruling a state.

---

[26] Democracy Rating 2015.Scores // http://democracyranking.org/ranking/2015/data/Scores_of_the_Democracy_Ranking_2015_A4.pdf (Accessed 16.09.2016)

[27] Global Democracy Ranking // http://democracyranking.org/wordpress/theoretical-basis/ (Accessed 16.09.2016)

[28] The Economist Intelligence Unit's Index of Democracy 2010// http://graphics.eiu.com/PDF/Democracy_Index_2010_web.pdf (Accessed 26.09.2016)

[29] The Economist Intelligence Unit's Index of Democracy 2010// http://graphics.eiu.com/PDF/Democracy_Index_2010_web.pdf (Accessed 26.09.2016)

[30] Cit.ex. Churchill on democracy // http://www.democracy-building.info/definition-democracy.html (Accessed 19.09.2016)

REFERENCES

- Amartya Kumar Sen. Democracy as a Universal Value // Journal of Democracy, July 1999 // http://www.journalofdemocracy.org/articles/tables-contents/July-1999
- Diamond L., Plattner Marc F., Chu Yun-Han. Democracy in East Asia: A New Century// The Johns Hopkins University Press, 1998.
- Doyle M.W. Liberal Peace: Selected Essays // New York: Routledge, 2011.
- Held D. Political Theory and the Modern State. Essays on State, Power, and Democracy // Stanford University Press, 1989.
- Khan Ali L. A Theory of Universal Democracy: Beyond the End of History // Martinus Nijhoff Publishers, 2003.
- Kochetkov V.P. Philosophy Of The Russian Constitution: Russian Values And Democracy // Sententia. European journal of humanities and social sciences, №2, 2013
- Kotze D., Nwonwu F. African Political Elites. The search for democracy and good governance // Africa Institute of South Africa, 2008.
- Linz J. J. and Stepan A. Problems of Democratic Transition and Consolidation: Southern Europe, South America, and Post-Communist Europe // JHU Press, 1996.
- Moore C. The European elite forgot that democracy is the one thing Britain holds most dear // The Telegraph, 24.06.2016 // http://www.telegraph.co.uk/news/2016/06/24/the-european-elite-forgot-that-democracy-is-the-one-thing-britai/
- Netanyahu: Israel Is a Democracy and Won't Tolerate Discrimination //Haaretz, 25.12.2011 // http://www.haaretz.com/israel-news/netanyahu-israel-is-a-democracy-and-won-t-tolerate-discrimination-1.403473
- Pourhiet A.-M. Le. Définir la démocratie // Rev. française de droit constitutionnel. – P., 2011.
- Ranciere J. Hatred democracy // Verso, 2006.
- Reiter D. Democratic Peace Theory // http://www.oxfordbibliographies.com/view/document/obo-9780199756223/obo-9780199756223-0014.xml
- Rosanvallon P. Counter-Democracy:Politics in an age of distrust// Cambridge University Press, 2008.
- Russet B. Grasping the Democratic Peace: Principles for a Post–Cold War World // Princeton University Press, 1993.
- Russett,B. and John R. Oneal. Triangulating Peace: Democracy, Interdependence, and International Organizations // New York: Norton, 2001.
- Weart S. R. Never at War: Why Democracies Will Never Fight One Another // Yale University Press, 1998.

Humanities, Environment, Social Affairs.

Getting Wiser with The Food We Consume: The Key to Diminish Global Famine and Climate Change.

Carlos Hugo Radic Biazik,
Postsecondary Student, Centennial College, Canada
Bachelor of Civil Engineering, UNEFA, Venezuela.

1. INTRODUCTION

The Food and Agriculture Organization (FAO) conducted in 2011 a study that claimed that worldwide more than 1,300,000,000 tons of food are lost or wasted every year[31]. Ironically, in the same year United Nations reported that nearly 900 million people did not have enough food to meet their daily 2000 Cal for a balanced diet, therefore, people were in the stage of undernourishment. We live in a world where food is being overproduced and it is wasted in scandalous amounts. Yet food security represents an urgent challenge for heads of states because of the exponential growth of the population and the hassle that global warming and climate change impose into the production.

Although, the term "food security" originated first in 1974 at the World Food Conference, it has constantly been revised and updated with the incorporation of societal issues and human rights. The most recent and widely accepted definition states that "it is a condition in which all people, at all times, have physical and economic access to sufficient, safe and nutritious food to meet their dietary needs and food preferences for an active and healthy life" (World Food Submit, 1996). However, this definition still seems disconnected from attributing responsibilities to the consumer. In other words, consumers are not yet accountable for their negligence and those that have already met their food requirements became responsible for assuring that the resources they are employing do not work against others trying to achieve the same goal.

So much is being done already to increase food production. Human in the history of humanity has never had such success in food production. Genetically modified crops, that leads to stronger and more resistant crops to the adversities of climate, enhanced nutritional values, longer life spans and others. And yet the question for some governments is how they can produce more food for everyone. In fact, the food being wasted represents the solution for tackling food insecurity. However, we have never been well educated on how to correctly handle the food we consume. And education on food waste reduction is the key to feed the almost 10 billion people in the world by 2050[32] .

This paper studies the implications of food waste at consumer's level, explore the causes that lead to this conduct and the consequences to the economy, the environment and the society. Subsequently, explores solutions and describes alternatives that could better address food security and diminish global famine.

2. FOOD LOSS VS FOOD WASTE

2.1 Food Loss

It is directly associated with the reduction in the quality of the food. In other words, food that spoils before reaches the consumer. The causes are usually unintentional, and they are the result of technical limitation, bad refrigeration, deficiencies in infrastructure, packaging, handling and other procedural problems in the production and distribution stages.

2.2 Food Waste

Defined as a good quality food suitable for human consumption, but ends up wasted because of retailers or consumer's negligence. In other words, food that meets or exceeds safety standards for consumption is thrown away on a conscious decision due to product imperfections (colour, size or shape), rigid or misinterpreted label rules, inappropriate storage, and purchasing behaviour or improper cooking practices.
It is important to mention that "food loss" usually is more predominant in developing countries, perhaps because economic limitations and technology deficiencies, in the other hand, "food waste" have a tendency to be more established in developed countries,

[31] Global Food Losses and Food Waste (2011). The results of the study suggest that roughly one-third of food produced for human consumption is lost or wasted globally, which amounts to about 1.3 billion tons per year.

[32] United Nations: Department of Economic and Social Affairs states that the World population is projected to reach 9.7 billion by 2050.

where food is aggressively marketed, and seen as merchandise or commodity rather than a basic need.

3. IMPACT OF FOOD WASTE

3.1 Economic Impact

In order to quantify the global economic impact is necessary to develop a proper measurement system, have access to unrestricted and un-manipulated data, and articulate a better understanding on how cultures and countries look at food waste differently, which means that waste for some cultures is not for others. For instance, evaluating food waste in North America, Neff et al. 2015, found that Americans and Canadians waste together close to $231 billion on uneaten food. Yet, that cost only represents 29% of the full cost in the economy. Cost of labour, transportation of the goods to the markets, or food left unharvested are not even accounted.

The economic losses from food waste are colossal. The monetary value employed could be better invested in the environment or funding social programs. For instance, all the money lost in North America due to food waste could cover most of Canada's federal budget. At the individual level, food wastage in homes adds about $700 to $1500 a year to a household's grocery bill. At the retail level, the value of unnecessary food waste is estimated at $165 billion annually, represents more than 20% of the $750 billion wasted worldwide, which is 700% the investment in development Aid.

Food waste also impacts on the cost of food as a direct consequence. The current economic system based on the principle of supply and demand creates an unsustainable model. Waste food induces the demand for "needing" more food, therefore, more resources are used to grow and transport food, like water and energy, which become costlier since they are more utilized. In other words, the more we "need" to grow food in an increasing population-environment make the cost higher. Those costs translate to higher costs at the production and distribution level, which customers end up paying.

3.2 Ecological Impact

Whichever the terms we refer to, either "Food loss" or "food waste", both has required enormous amounts of energy, land, water, time, natural resources, human labour, chemicals, transportation, and others in order to be produced. All this process creates CO2 emissions that contribute to the global warming which threaten the efforts for food security. In other words, food waste becomes a problem that requires more resources to be treated.

3.2.1 CO2 footprint of global food wastage

According to the FAO, food waste produces 3.3 gigatons of CO2 and represents the third largest source of CO2 emissions in the world . Furthermore, the food that is produced and subsequently discarded, occupies approximately 30% of the farming land in the world impacting also on the biodiversity due to the harvesting practices and deforestation. Figure 1 Figure 1 shows the CO2 emissions of 15 of the most CO2 emitter countries in the world and adds food wastage like as if this were a country, so, it would occupy be the third largest CO2 emitters after China and USA. shows the CO2 emissions of 15 of the most CO2 emitter countries in the world and adds food wastage like as if this were a country, so, it would occupy be the third largest CO2 emitters after China and USA.

Figure 1. Total CO2 emissions of top 15 of countries (year 2011) vs. Food wastage.

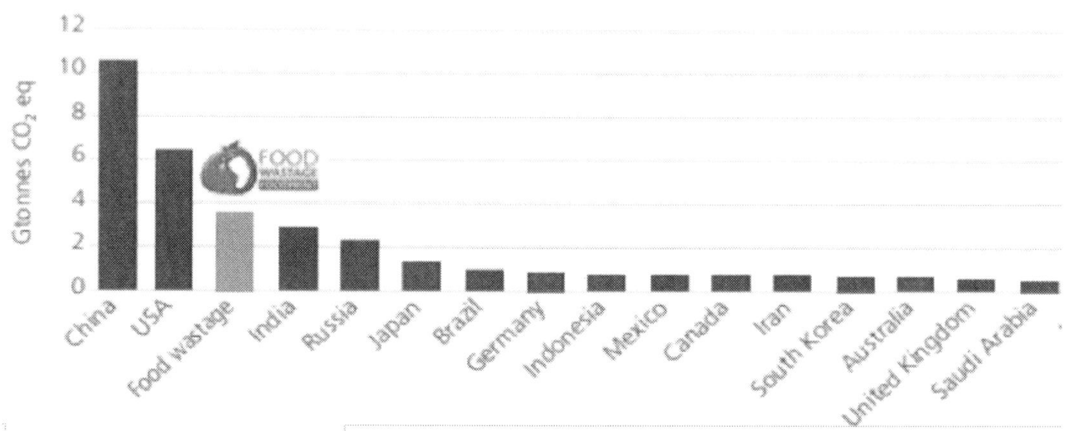

Source: Food Wastage Footprint Full-cost Accounting (2014).

3.2.2 CO2 Emissions by Product

The associated CO2 emissions of food waste depend on fundamentally 3 main factors, first, the location where the food is being produced; second, the type of product and; third the stage in which the food is being discarded. The location where vegetable production is being produced impacts on the Carbon emissions generated. For instance, Europe is more carbon intensive than vegetable production in Industrialized and Southeast Asia, as Europe uses more carbon-intensive means of production, such as artificially heated greenhouses. Inversely, cereal production in Asia is more carbon intensive than cereal production in Europe due to the difference in the type of cereal grown: rice on average has higher impact factors than wheat. Rice is a CH4 emitting crop because of the decomposition of organic matter in paddy fields (1 kg of CH4 is the equivalent of 25 kg of CO2).

Figure 2. The contribution of each phase of the food supply chain to carbon footprint and food wastage.

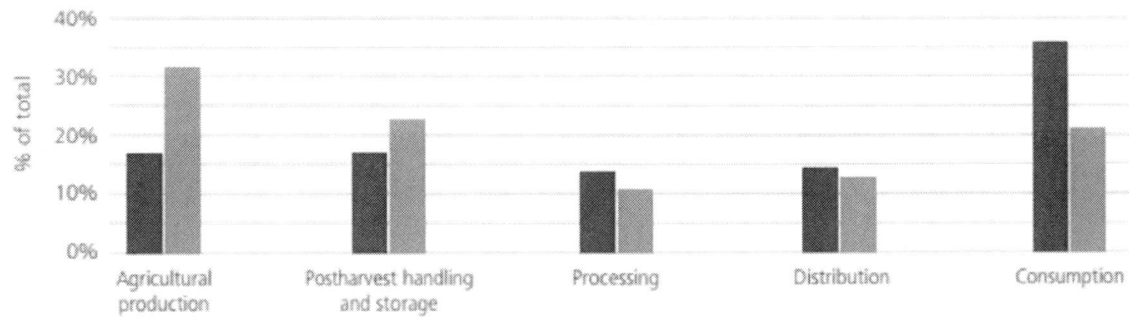

Source: Food Wastage Footprint Full-cost Accounting (2014).

3.3 Ethical and Social Impact

In addition to the economic and environmental costs, food wastage represents a variety of social and ethical costs. The environmental costs of carbon emissions, which were described in section 3.2.1 also transform in social costs. For instance, agriculture activities lead to added costs in individual and societal health costs due to resource depletion and pollution. In other words, food wastage generates food security risks, loss of habitat and landscape that threaten the fauna, loss of livelihoods and possibilities of social conflicts, crime and violence due to resource depletion and lack of equitable access to those.

Figure 3. The current world population of 7.3 billion is expected to reach 8.5 billion by 2030, 9.7 billion in 2050 and 11.2 billion in 2100.

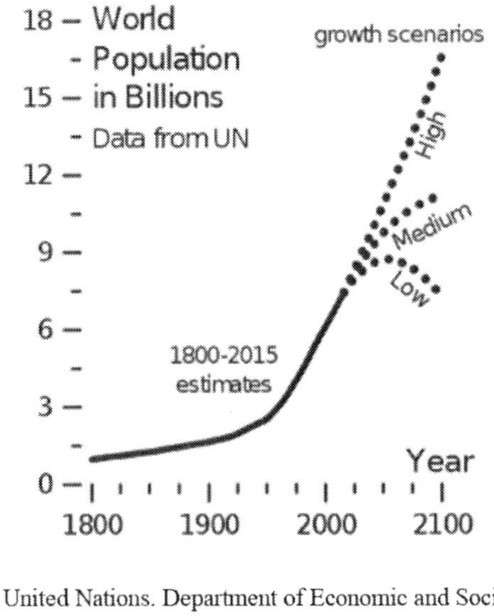

Source: United Nations. Department of Economic and Social Affairs.

Moreover, the actual production system of food is ethically questionable. Farmers and local producers enter into a market where their food is treated as a merchandise, so their effort and the value of their products are frequently dishonoured. Likewise, governments commit to deals that create barriers for locals to access land, and natural resources for their own production. Priority is given to the products that would benefit the country economically or the produce that will feed the animals for the meat industry. Currently, the global production of food could feed as much as 12 billion people, however, 40% of the world food production is wasted, creating a societal problem that leads to 1.3 billion tons of food being wasted where almost 900 million undernourished people.

4. ARE CONSUMERS TO BLAME FOR ALL THE FOOD IS WASTED?

According to Maus et. Al (1988), there are four changes in the household behaviours towards food consumption, that represent part of the explanation for food waste. The authors highlight first the "devaluation" of the food significance, which is seen as a product and not as need; second and third disinformation about "food identity" and "food origin", in other words, the disconnection of cultural background of the food they access such as the ingredients and dishes history, loss of local food culture being globalization one of the main attributable reasons; and last "loss of social and emotional linkage to food", where the authors explain that siting in Family to eat together is not longer a tradition. Therefore, these four changes in the household behaviours have lead to an emotional disconnection to the food we consume and add the wrong perception that food is a product instead of something of vital need.

Unfortunately, in the global economy consumers are encouraged to purchase more food that they really need thanks to severe and vicious sells practices. This is exactly the behaviour must change if we do want to reduce the food that is being wasted. Whether or not changes in societal values are the causes of consumers disconnection to the value of the food they acquire, consumers are still responsible for their actions on food waste.

Figure 4. Per capita food losses and waste (Kg/year), at consumption and pre-consumption stages, in different regions of the world.

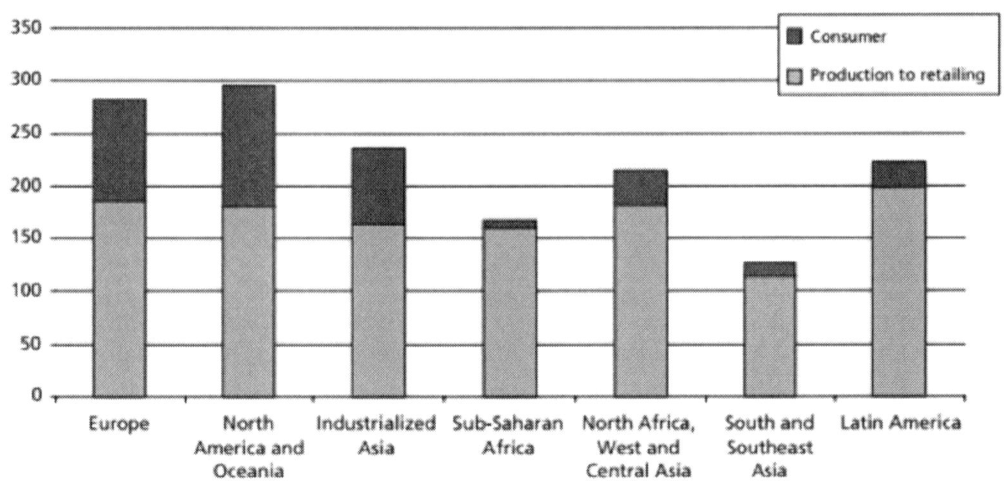

Source: Food Wastage Footprint Full-cost Accounting (2014)

Per capita Food losses and food waste at production and consumption levels are sown in figure 4. The graph displays seven different regions against the amount of food being wasted. Furthermore, the analysis of the graph can be interpreted by breaking down the data into developed countries (Europe, North America and industrialized Asia) and developing countries (Africa, South Asia and Latin America). It could be seen that the food wasted at production level is fundamentally the same no matter the region we are looking at; however, considerable differences are found when evaluating the food waste at the consumption level. First, the waste produced in developing countries at consumption level represents not more than 15% in the worst of the cases in comparison with the waste at the production level. Second, the total waste at the consumption level of the fourth developing regions (combined) is approximatively similar to the same item in North America. Per capita food losses in North America and Europe are approximately 300 kg/yr. each, conversely, in Southeast Asia, North sub-Saharan Africa, the production is nearly 170 Kg/year each. In North America are produced approximately 900 kg/yr. of food, similar numbers for Europe, in South Asia approximately 460 kg/yr. and 200 Kg/Year in Latin America. Regrettably, 95 to 115 kg /yr of the food produced in North America and Europe are lost at consumption level. In other regions, the losses only account for 6 to 11 kg /year, which represents without any doubt a different approach to the importance given to food by the consumers.

From this numbers, we could agree that all the food waste in North America could totally replace all the food loss at a production level of South and Southeast Asia. Similarly, adding the waste of food of Europe and Industrialized Asia would be enough to cover the losses at a production level of Latin-America.

5.CONCLUSIONSANDRECOMMENDATIONS

Certainly, consumers are the last piece in the equation on food waste, nonetheless, they are the ones that with proper education can contribute the most to the problem. Stopping and reducing consumer waste of food could advance food security, significantly improve nutrition and diminish the negative effects on the environment and impact on public health. Changes in food waste policies, business practices and individual incentives to reduce waste must be formed. We have to create campaigns that discourage the compulsory consumption behaviour of purchasing, provide education towards food freshness, expiration or due date labeling standards; create incentives for restaurants to provide smaller portions, implementing a waste management payment system (or in other words "fines" for waste food) according to the weight of household waste produced, better regulate sells at retailer's level, as well as, the weekly, monthly promotional and loyalty programs, etc. Those are some of the examples that can be implemented to tackle the problem that keeps almost 900 million undernourished people.

The possible solutions to reduce food waste can be grouped in 2 categories, immediate solutions, that although are not sustainable, are needed in the short term and sustainable solutions that in other words will not create other problems.

Table 1. Group of solutions for food waste

5.1 Immediate Solutions

- Secondary resellers. (Sell after expiration dates).
- Home Composting.
- In-vessel Composting.
- Community Composting.
- Water resources management facilities.
- Centralized anaerobic digestion.
- Animal feed.
- Tax donation incentives.
- Apply fines.

5.2 Sustainable Solutions

- Food waste reduction initiatives.
- Consumers to buy and prepare food with a plan.
- Food recycling.
- Better labelling, better Understanding.
- Smaller plates, smaller portions.
- Consumption of imperfect produce (wrong shape, colour, size, etc).
- "Freshness", non-fresh does not mean toxic.
- Understanding retail marketing campaigns.

Finally, solutions for solving the problem of food waste resides on changing consumers 'patterns of behaviour through proper education. The paper focused on solutions that pass by changes in food waste policies, business practices and individual incentives to reduce waste. It was stressed that the relevance of looking into the dynamics of demand and supply to tackle food waste and that the solutions for this do not necessarily overlap with solutions to famine in developing countries. Added research would be relevant to improve the distribution of food to more deprived areas. In areas with food shortage, famine eradication crucially depends on the design of institutional changes leveraged by technological development.

REFERENCES

- Caicedo, N. B. M., & Ibarra, A. A. R. (2017, August). Estado actual de los niveles de desperdicio de las cadenas de abastecimiento de alimentos. In Memorias de Congresos UTP (pp. 202-209).
- Food Waste and Consumer Ethics (PDF Download Available). Available from: https://www.researchgate.net/publication/264976655_Food_Waste_and_Consumer_Ethics [accessed Dec 1 2017].
- Fox, T., & Fimeche, C. (2013). Global food: waste not, want not. London: Institution of Mechanical Engineers.
- Gjerris, M., & Gaiani, S. (2014). Food waste and consumer ethics. In Encyclopedia of Food and Agricultural Ethics (pp. 1000-1006). Springer Netherlands.
- Godfray, H. C. J., Beddington, J. R., Crute, I. R., Haddad, L., Lawrence, D., Muir, J. F., ... & Toulmin, C. (2010). Food security: the challenge of feeding 9 billion people. science, 327(5967), 812-818.
- Gunders, D. (2012). Wasted: How America is losing up to 40 percent of its food from farm to fork to landfill. Natural Resources Defense Council, 26.
- Gustavsson, J., Cederberg, C., Sonesson, U., Van Otterdijk, R., & Meybeck, A. (2011). Global food losses and food waste (pp. 1-38). Rome: FAO.
- Gustafsson, J., Cederberg, C., Sonesson, U., & Emanuelsson, A. (2013). The methodology of the FAO study: Global Food Losses and Food Waste-extent, causes and prevention"-FAO, 2011.
- Johnston, J., Szabo, M., & Rodney, A. (2011). Good food, good people: Understanding the cultural repertoire of ethical eating. Journal of consumer culture, 11(3), 293-318.
- Kummu, M., de Moel, H., Porkka, M., Siebert, S., Varis, O. & Ward, P.J. 2012. Lost food, Wasted Resources: Global Food Supply Chain Losses and their Impacts on Freshwater, Cropland, and Fertiliser Use. The science of the Total Environment, 438, pp.477–489.
- Lipinski, B., Hanson, C., Lomax, J., Kitinoja, L., Waite, R., & Searchinger, T. (2013). 10 Ways to Cut Global Food Loss and Waste. World Resources Institute Working Paper, June.
- (2013). Reducing food loss and waste. World Resources Institute Working Paper, June.
- Maus, N., Paul, T., Pudel, V., & Westenhöfer, J. (1988). Behavioural factors involved in the control of food intake in man. International journal for vitamin and nutrition research. Internationale Zeitschrift fur Vitamin-und Ernahrungsforschung. Journal international de vitaminologie et de nutrition, 58(3), 356-366.
- Neff, R. A., Spiker, M. L., & Truant, P. L. (2015). Wasted food: US consumers' reported awareness, attitudes, and behaviours. PloS one, 10(6), e0127881.
- Pantsios, A. (2015). Reducing Food Waste Is Good for Economy and Climate. Ecowatch. Available from https://www.ecowatch.com/reducing-food-waste-is-good-for-economy-and-climate-report-says-1882013448.html. [accessed Dec 1 2017].
- Pinstrup-Andersen, P. (2009). Food security: definition and measurement. Food security, 1(1), 5-7.
- Schott, A. B. S., Aspegren, H., Bissmont, M., & la Cour Jansen, J. (2013). Modern Solid Waste Management in Practice: The City of Malmö Experience. Springer Science & Business Media.
- Stuart, T. 2009. Waste – Uncovering the Global Food Scandal. Penguin Books: London, ISBN: 978-0-141-03634-2.

WORKS CITED

- Food and Agriculture Organization of the United Nations (2014). Food Wastage Footprint Full-cost Accounting: Final Report. Food and Agriculture Organization (Fao). Available from http://www.fao.org/3/a-i3991e.pdf. [accessed Dec 1 2017].
- Gjerris, M., & Gaiani, S. (2014). Food waste and consumer ethics. In Encyclopedia of Food and Agricultural Ethics (pp. 1000-1006). Springer Netherlands.
- Gustavsson, J., Cederberg, C., Sonesson, U., Van Otterdijk, R., & Meybeck, A. (2011). Global food losses and food waste (pp. 1-38). Rome: FAO.
- Lipinski, B., Hanson, C., Lomax, J., Kitinoja, L., Waite, R., & Searchinger, T. (2013). 10 Ways to Cut Global Food Loss and Waste. World Resources Institute Working Paper, June.
- Maus, N., Paul, T., Pudel, V., & Westenhöfer, J. (1988). Behavioural factors involved in the control of food intake in man. International journal for vitamin and nutrition research. Internationale Zeitschrift fur Vitamin-und Ernahrungsforschung. Journal international de vitaminologie et de nutrition, 58(3), 356-366.
- Miles, R. (2009). ICT Applications for Sustainable Development Focusing on Climate Change Preparedness: A discussion paper prepared for the United nations department of economic and social affairs. Avaiable from http://unpan1.un.org/intradoc/groups/public/documents/UN-DPADM/UNPAN037120.pdf. [accessed Nov 1 2017].
- Weber, C. L., & Matthews, H. S. (2008). Quantifying the global and distributional aspects of American household carbon footprint. Ecological economics, 66(2-3), 379-391.
- World Food Submit. (1996). Rome Declaration on World Food Security and World Food Summit Plan of Action. Available from: http://www.fao.org. [accessed Dec 1 2017].

Research of Game Strategy on the Online Public Opinion in the Mode of Individual Help for Online Donation

Lin Zhang,
MSc,Tianjin University of Finance and Economics,China

1. INTRODUCTION

In the era of «Internet +», the public participation rate of charity activities has been greatly improved by the online donation, as a new way of charitable donations, online donation has been more and more accepted among the public. However, the anonymous and private problem of the new model can not guarantee the authenticity of the donation information,we take Tianjin bombings event as an example, Sina Microblogging platform users called «my heart belong to Bayern Munich always» utilized the negative impact of the incident, falsely claimed that his father was killed in the Tianjin bombings to deceive the reward;» know the goddess «Tong Yao fraud on the platform with the Kochi users' representative group and the «Rolle incident» broke out on the WeChat platform at the end of 2016, eventually , individual help was evolved into marketing tools,consumed donators' trust and love, such events triggered online public opinion pushing network donation to the cusp. Standardized individual help is to seek fundraising by virtue of charitable organizations. In reality, most of the individual users prefer to seek help in the WeChat friends circle and other social platform through the acquaintance of mutual forwarding and other donations, this way of transforming rescue information formed a «information aggregation», promoted to online public opinion (FANG Fu-jian 2011) developing. However,individual help fundraising was easy to be malicious abuse of the request without a rigorous review and supervision, and published fictitious information by virtual of public good faith, once the event exposured, the public interest would be damaged as well as the tendency of public opinion would be difficult to control(Shibo 2010), resultly,it can not only intensified the online public opinion's fermentation, but also reduced the credibility of the network public welfare. Therefore, the government should provide guidance on the information,as well as take positive measures to deal with the outbreak of online public opinion, under the rational control of cyberspace and the appropriate punishment for the subject, to compensate for public interest,and to make the negative online public opinion be reasonable soluted.

2. LITERATURE REVIEW

Online public opinion refers to netizens with different views on the social issues expressing and disseminating a strong influence on the collection of emotional, public opinion and behavior's tendencies through the Internet media(Zeng Runxi 2009). The evolution of the online public opinion is the result of the interaction of the participants, participants can choose the appropriate behavior strategy, take the initiative to promote the online public opinion to the expected direction(Bin Ning, Yank Shubiao & Hu Feng 2017). At present, research scholars for the evolution of online public opinion are mostly from the perspective of the incident itself, government's behavior and public opinion. The overseas research on the evolution of online public opinion mainly focuses on explaining the evolution process of public opinion through the particle interaction model, such as Sznajd model and its modified model(Sznajd J N K 2000), limited liability model(Weisbuch G 2004). In addition, such twitter, microblog and other social platforms as an important media for the spread of online public opinion due to less information, widely disseminated and other prominent features become the focus of research(WANG Shu-xian 2016). While domestic scholars focus on the evolution of online public opinion and influence factors. Sun Yixiao (2016)considered that some of the opinion leaders have significant effects but lack a sense of self-discipline which based on the matrix model analysis. Wang Xu et al(2017) took «Wei Zexi event» as an example, combined with social network analysis and quantitative methods to explore the process of online public opinion's diffusion. Qi Kai et al(2017) took evolutionary theory of the game into the analysis of the online public opinion transmission of the netizen groups, and found that the punishment mechanism and the initial state of the netizen played an important role in the guidance and extendency of public opinion. Zhang Lifan et al(2017) constructed the utility function game matrix with fairness factors to analyze the selection of network media-government, netizen-opinion leaders in the subjective game process, to obtain the perfect subgame equilibrium path conditions. This paper takes the «Rohr event» as an example, and creatively introduces the dynamic evolutionary game method to analyze the evolution process of the network public opinion in the model of individual help for online donations,and puts forward the guidance and recommendations for multi-participant subject's strategies selection in the different stages of public opinion, and thus effectively inhibit the negative impact of online public opinion.

3. ANALYSIS OF THE GAME EQUILIBRIUM OF PUBLIC OPINION EVOLUTION IN THE MODE OF INDIVIDUAL HELP FOR ONLINE DONATION

3.1 Research on the Formation of Online Public Opinion in the Mode of Individual Help for Online Donation

Collections of participants' strategy:

The selection set of the recipients' strategy is{publish the real information, publish the fictitious information}; the selection set of business' strategy is {cooperate, no cooperate}; the selection set of netizens' strategy is {donor, no donor}. Donations here are rewarded, forwarded or directed. Table 1 shows the recipients-business-netizens' income risk matrix.

Table 1.Recipients-Business-Netizens Income Risk Matrix

Strategy combination	recipients	business	netizens
(publish the real information,cooperate,donor)	U1+α1-u1-a1-a2	W1+θ1-w1	R1+σ1-r1+R2+R2+W2
(publish the real information,cooperate,no donor)	U1-u1-a1-b1	W1-w1	0
(publish the real information,no cooperate,donor)	U1+β1-u1-a2	-c2	R1+σ1-r1+R2
(publish the real information,no cooperate,no donor)	U1-u1-b1	0	0
(publish the fictitious information,cooperate,donor)	U1+α1-u1-a1-a2-b3	W1-w1-c1	R1-r1-r2-w2
(publish the fictitious information,cooperate,no donor)	U1-u1-a1	W1-w1-c1	σ2-w2
(publish the fictitious information,no cooperate,donor)	U1+β1-u1-a2-b2	0	R1-r1-r2
(publish the fictitious information,no cooperate,no donor)	U1-u1	θ3	σ2+W2

Table 2 shows the related risk parameters involved in the income risk matrix.

Table 2.Parameter Definition

symbol	Symbol Description
U1	Recipients can obtain normal income through the social platform for help.
α1	Recipients can obtain additional benefits from netizens with business assistance .
β1	Recipients can't obtain additional benefits from netizens without business assistance .
u1	Recipients need to pay the cost through the online social platform for help.
a1	Recipients need to pay the cost for enterprises' funding.
a2	Recipients need to pay the cost for getting the netizen's trust and donations.
b1	Recipients need to take risks of difficult raising funds and delaying illness when they published the real information without netizen's trust and donation .
b2	Recipients need to take risks of the pressure from donors' condemn and denounce when they published the fictitious information .
b3	Recipients need to take risks of accepting more serious penalties for cooperating with companies with publishing the fictitious information.
W1	Enterprises can receive brand revenue through participating in public welfare.
W2	Enterprises can bring positive and stable benifits with promoting the development of online public welfare to the community .
θ1	Enterprises can receive additional benefits from netizen's donation by assisting in the issuance of recipients who published real information.
θ2	Enterprise can protect the positive image for colliding with the recipients who published the fictitious information .
w1	Enterprises and recipients'cooperation need to pay the cost of network fundraising activities.

c1	Enterprises can be condemned by the public and bring in other negative impact because of colliding with recipients who pubished the fictitious information .
w2	Enterprises hinder the development of Internet public welfare,and are questioned by the public as well as bring in other negative impact because of colliding with recipients who pubished the fictitious information .
c2	Enterprise has lost the opportunity to establish a brand image because it has not assisted the person who published the real information at the same time obtained the donation from the netizens.
R1	Netizens receive social benefits from recipients who published the real information .
σ1	Netizens receive the additional benefits from donating recipients who published the real information..
σ2	Netizens receive the additional benefits from not donating recipients who published the fictitious information.
r1	Netizens need to pay the cost through the social platform for donor.
r2	Netizens could lose money and have a sense of psychological loss and other discontent due to donate recipients who published the fictitious information.
R2	Netizens promote the orderly and positive development of Internet public welfare due to donate recipients who published the real information.

Assuming that the proportion of recipients choose to release the proportion of real information is x, then select the ratio of releasing the fictitious information (1-x);the proportion of enterprises choosing to cooperate with the recipients is y, and the proportion of non-cooperation is chosen as (1-y); the proportion of netizens choose to donate is z, choose not to donate the ratio of (1-z);
The recipients choose the benefits of different strategies available as:

$$A_{11} = yz(U_1+\alpha_1-u_1-a_1-a_2)+y(1-z)(U_1-a_1-u_1-b_1)+(1-y)z(U_1+\beta_1-u_1-a_2)+(1-y)(1-z)(U_1-u_1-b_1)$$

$$A_{12} = yz(U_1+\alpha_1-u_1-a_1-a_2-b_3)+y(1-z)(U_1-a_1-u_1)+(1-y)z(U_1+\beta_1-u_1-a_2-b_2)+(1-y)(1-z)(U_1-u_1)$$

$$\overline{A_1} = xA_{11}+(1-x)A_{12}$$

So the recipients choose to publish the real information strategy with the dynamic replication equation can be referred as follows:

$$F(x)=\frac{dx}{dt}=x(1-x)(A_{11}-\overline{A_1})=x(1-x)(-b_1+zb_1+zb_2-yzb_2+yzb_3) \quad (1)$$

Make $F(x)=0$,so the balance of the three-way game can be obtained $x=1,0, y=\frac{b_1}{b_3-b_2}\frac{1}{z}-\frac{b_1+b_2}{b_3-b_2}$,

when $y=\frac{b_1}{b_3-b_2}\frac{1}{z}-\frac{b_1+b_2}{b_3-b_2}$, at this time $F(x)\equiv 0$; regardless of what value of x, tripartite interests can get the

maximum, all points are dynamic equilibrium points; when $y>\frac{b_1}{b_3-b_2}\frac{1}{z}-\frac{b_1+b_2}{b_3-b_2}$,

$\left.\frac{\partial F(x)}{\partial x}\right|_{x=1}<0, \left.\frac{\partial F(x)}{\partial x}\right|_{x=0}>0$ at this time $x=1$ is the equilibrium point, then the recipients will choose to

publish the real information; when $y<\frac{b_1}{b_3-b_2}\frac{1}{z}-\frac{b_1+b_2}{b_3-b_2}$, because $\left.\frac{\partial F(x)}{\partial x}\right|_{x=0}<0, \left.\frac{\partial F(x)}{\partial x}\right|_{x=1}>0$, at that

time $x=0$, is the equilibrium point,thus the recipients will choose to publish fictitious information.
Similarly, enterprises choose the benefits of different strategies available as:

$$B_{11} = xz(W_1+\theta_1-w_1)+x(1-z)(W_1-w_1)+(1-x)z(W_1-w_1-c_1)+(1-x)(1-z)(W_1-w_1-c_1)$$

$$B_{12} = xz(-c_2)+(1-x)z\theta_2$$

$$\overline{B_1} = yB_{11}+(1-y)B_{12}$$

the dynamic replication equation can be referred as follows:

$$F(y)=\frac{dy}{dt}=y(1-y)(B_{11}-B_{12})=y(1-y)(xz\theta_1+W_1-w_1-c_1+xc_1+xzc_2-\theta_2+x\theta_2) \quad (2)$$

Make So the balance of the tripartite game can be obtained $y=1,0, z=\frac{-W_1+w_1+c_1+\theta_2}{\theta_1+c_2}\frac{1}{x}-\frac{c_1+\theta_2}{\theta_1+c_2}$, when $z=\frac{-W_1+w_1+c_1+\theta_2}{\theta_1+c_2}\frac{1}{x}-\frac{c_1+\theta_2}{\theta_1+c_2}$,at this time $F(y)\equiv 0$,regardless of what value of y, tripartite interests can get the maximum, all points are dynamic equilibrium points;

when $z>\frac{-W_1+w_1+c_1+\theta_2}{\theta_1+c_2}\frac{1}{x}-\frac{c_1+\theta_2}{\theta_1+c_2}$, because $\left.\frac{\partial F(y)}{\partial y}\right|_{y=1}<0,\left.\frac{\partial F(y)}{\partial y}\right|_{y=0}>0$, at this time $y=1$ is the equilibrium point, then enterprises will choose to cooperate with the recipients;

when $z<\frac{-W_1+w_1+c_1+\theta_2}{\theta_1+c_2}\frac{1}{x}-\frac{c_1+\theta_2}{\theta_1+c_2}$, because $\left.\frac{\partial F(y)}{\partial y}\right|_{y=0}<0,\left.\frac{\partial F(y)}{\partial y}\right|_{y=1}>0$, at this time $y=0$ is the equilibrium point, then enterprises will refuse to cooperate with the recipients; Thus netizens choose the benefits of different strategies available as:

$$C_{11}=xy(R_1+\sigma_1-r_1+R_2+W_2)+x(1-y)(R_1+\sigma_1-r_1+R_2)+(1-x)y(R_1-r_1-r_2-w_2)+(1-x)(1-y)(R_1-r_1-r_2)$$
$$C_{12}=(1-x)y(\sigma_2-w_2)+(1-x)(1-y)(\sigma_2+W_2)$$
$$\overline{C_1}=zC_{11}+(1-z)C_{12}$$

Netizens choose to donate with the dynamic replication equation can be referred as follows:

$$F(z)=\frac{dz}{dt}=z(1-z)(C_{11}-C_{12})=z(1-z)(x\sigma_1+R_1-r_1-r_2+xR_2+x\sigma_2-\sigma_2+xr_2+yW_2+xW_2-W_2) \quad (3)$$

Make $F(z)=0$,so the balance of the tripartite game can be obtained

$$z=1,0,x=-\frac{W_2}{W_2+\sigma_1+R_2+r_2+\sigma_2}y+\frac{W_2+r_1-R_1+r_2+\sigma_2}{W_2+\sigma_1+R_2+r_2+\sigma_2},$$

when $x=-\frac{W_2}{W_2+\sigma_1+R_2+r_2+\sigma_2}y+\frac{W_2+r_1-R_1+r_2+\sigma_2}{W_2+\sigma_1+R_2+r_2+\sigma_2}$, $F(z)\equiv 0$,

at this time regardless of what value of z, tripartite interests can get the maximum, all points are dynamic equilibrium points; when $x>-\frac{W_2}{W_2+\sigma_1+R_2+r_2+\sigma_2}y+\frac{W_2+r_1-R_1+r_2+\sigma_2}{W_2+\sigma_1+R_2+r_2+\sigma_2}$,

because $\left.\frac{\partial F(z)}{\partial z}\right|_{z=1}<0,\left.\frac{\partial F(z)}{\partial z}\right|_{z=0}>0$ at this time $z=1$ is the equilibrium point, then netizens will choose to donate; when $x<-\frac{W_2}{W_2+\sigma_1+R_2+r_2+\sigma_2}y+\frac{W_2+r_1-R_1+r_2+\sigma_2}{W_2+\sigma_1+R_2+r_2+\sigma_2}$,because $\left.\frac{\partial F(z)}{\partial z}\right|_{z=1}>0,\left.\frac{\partial F(z)}{\partial z}\right|_{z=0}<0$

at this time $z=0$ is the equilibrium point, then netizens will choose not to donate.

Based on the replicated dynamic equation method, the equilibrium point analysis can obtain the equilibrium point of the local steady state of the online public opinion in the formation phase. Therefore, the local stability of each equilibrium point can be analyzed by using the Jacobian matrix to find the steady state of the online public opinion stability point. The Jacobian matrix is:

$$J=\begin{pmatrix} (1-2x)(-b_1+zb_1+zb_2-yzb_2+yzb_3) & (x-x^2)(-zb_2+zb_3) & (x-x^2)(b_1+b_2-yb_2+yb_3) \\ (y-y^2)(z\theta_1+c_1+zc_2+\theta_2) & \begin{matrix}(1-2y)(xz\theta_1+W_1-w_1-c_1\\+xc_1+xzc_2-\theta_2+x\theta_2)\end{matrix} & (y-y^2)(x\theta_1+xc_2) \\ (z-z^2)(\sigma_1+R_2+\sigma_2+r_2+W_2) & (z-z^2)W_2 & \begin{matrix}(1-2z)(x\sigma_1+R_1-r_1-r_2+xR_2+x\sigma_2\\-\sigma_2+xr_2+yW_2+xW_2-W_2)\end{matrix} \end{pmatrix}$$

Make $A = -b_1 + zb_1 + zb_2 - yzb_2 + yzb_3, B = xz\theta_1 + W_1 - w_1 - c_1 + xc_1 + xzc_2 - \theta_2 + x\theta_2,$
$C = x\sigma_1 + R_1 - r_1 - r_2 + xR_2 + x\sigma_2 - \sigma_2 + xr_2 + yW_2 + xW_2 - W_2$

$$TrJ = (1\text{-}2x)A + (1-2y)B + (1-2z)C$$

$$\begin{aligned}\det J &= (1-2x)(1-2y)(1-2z)ABC + xyz(1-x)(1-y)(1-z)[(-zb_2+zb_3)(x\theta_1+xc_2)(\sigma_1+R_2+\sigma_2+r_2+W_2)\\ &+(b_1+b_2-yb_2+yb_3)(z\theta_1+c_1+zc_2+\theta_2)W_2]-xy(1-x)(1-y)(1-2z)C(-zb_2+zb_3)(z\theta_1+c_1+zc_2+\theta_2)\\ &-yz(1-y)(1-z)(1-2x)A(x\theta_1+xc_2)W_2-xz(1-x)(1-z)(1-2y)B(b_1+b_2-yb_2+yb_3)(\sigma_1+R_2+\sigma_2+r_2+W_2)\end{aligned} \quad (4)$$

According to the results of the equilibrium analysis, we can see that when the enterprises choose the cooperative strategy as $y=1$,plus $W_1 - w_1 + \theta_1 > 0$, similarly when the user chooses the donation strategy as $z=1$,plus $R_1 + \sigma_1 - r_1 + R_2 + W_2 > 0$,which can be immediately and substituted into the formula (4). then we can find the equilibrium point is $(1,1,1),(1,0,1),(1,0,0),(0,0,0)$, and $\det J < 0$, the result indicate that the determinant is the saddle point and is locally stable state, thus the choice of the above strategy can not make public opinion to achieve the evolution of stable state, when participants' strategy affected by the outside world, they would tend to reverse the direction to make the overall stability of the fluctuations, and thus not enough to make the formation of public opinion continue to advance; when the equilibrium point is $(0,1,0),(0,0,1),(1,1,0)$ .and $\det J > 0$, TrJ uncertainty ,so the point is still saddle point, only in the local stable state, the above strategy is not evolutionary stability strategy; when the equilibrium point is $(0,1,1)$ meet the conditions that then tripartite interests at this time to achieve a balanced and stable state, thus the equilibrium point for the evolution Stability point is $(0,1,1)$, indicating that the recipients choose to publish fictitious information, companies choose cooperate and Netizens choose to believe and forward the strategy as the evolution of stability strategy.

3.2. Research on the Development of Online Public Opinion in the Mode of Individual Help for Online Donation

Collections of participants' strategy:
The selection set of the netizens' strategy is{believe, question}; when the netizens question the authenticity of help information from recipients will accelerate the spread of public opinion, otherwise, public opinion will be blocked. Table 3 shows the income risk matrix between netizens and netizens.

Table 3.Netizens-Netizens Income Risk Matrix

Strategy combination		netizens: believe	netizens: question
netizens	believe	U-a,U-a	R+β1|R-r|-(M+β1|M-m|),r+β2|R-r|-(m+β2|M-m|)
	question	r+β2|R-r|-(m+β2|M-m|),R+β1|R-r|-(M+β1|M-m|)	W-b,W-b

Table 4 shows the related risk parameters involved in the income risk matrix.

Table 4.Parameter Definition

symbol	Symbol Description
U	netizens A and netizens B obtained the psychological identity and other benefits for believing that the authenticity of the information from recipients
a	netizens are responsible for the risk of being deceived by the help of recipients

b	netizens need to pay the cost of public information dissemination for questioning.
W	netizens A and netizens B obtained the psychological satisfaction and other benefits for questioning the authenticity of help information from recipients.
R	When one of the netizens choose to believe the other choose to question, the believe one receive the inherent benefit.
r	When one of the netizens choose to believe the other one choose to question, the question one receive the inherent benefit.
M	When one of the netizens choose to believe the other one choose to question, the believe one take the risk.
m	When one of the netizens choose to believe the other one choose to question, the question one take the risk.
β1	The additional revenue and cost factor of the trust party available when the two parties did not reach the agreement.
β2	The additional revenue and cost factor of the question party available when the two parties did not reach the agreement.

Assuming that the proportion of netizens choose to believe is x, then select the ratio of questioning (1-x);Therefore, the netizens choose the benefits of different strategies available as:

$$A_{11} = (U-a)x + [R + \beta_1|R-r| - (M + \beta_1|M-m|)](1-x)$$

$$A_{12} = [r + \beta_2|R-r| - (m + \beta_2|M-m|)]x + (W-b)(1-x)$$

$$\overline{A_1} = xA_{11} + (1-x)A_{12}$$

The netizens choose to believe with the dynamic replication equation can be referred as follows:

$$F(x) = \frac{dx}{dt} = x(1-x)\{[U - a - (r + \beta_2|R-r|) + (m + \beta_2|M-m|)]x + (1-x)[-(W-b) + (R + \beta_1|R-r|) - (M + \beta_1|M-m|)]\} \quad (5)$$

Make $F(x)=0$ $A = U - a - (r + \beta_2|R-r|) + (m + \beta_2|M-m|)$ $B = -(W-b) + (R + \beta_1|R-r|) - (M + \beta_1|M-m|)$

We can get the equilibrium point of the three-way game as follows: $x = 1, 0, x^* = \frac{-B}{A-B}$ when $A>0, B>0$, if $A>B$, then $x^*<0$, else if $A<B$, then $x^*>1$, x is not the equilibrium point; at the same time $\left.\frac{\partial F(x)}{\partial x}\right|_{x=1} < 0, \left.\frac{\partial F(x)}{\partial x}\right|_{x=0} > 0$, $x=1$ is the only equilibrium point; when $A>0, B<0$, if $A>|B|$, then $0<x^*<1$, at the same time $\left.\frac{\partial F(x)}{\partial x}\right|_{x=1} < 0, \left.\frac{\partial F(x)}{\partial x}\right|_{x=0} < 0$, $x = 1, 0, x^*$ are the equilibrium points; if $A<|B|$,then $x^*<0$, thus x^* is not the equilibrium point, at the same time $\left.\frac{\partial F(x)}{\partial x}\right|_{x=1} < 0, \left.\frac{\partial F(x)}{\partial x}\right|_{x=0} < 0$, so there are two equilibrium points, respectively $x = 1, 0$; when $A<0, B>0$, then $0<x^*<1$ and $\left.\frac{\partial F(x)}{\partial x}\right|_{x=1} > 0, \left.\frac{\partial F(x)}{\partial x}\right|_{x=0} > 0$, x^* is the only equilibrium point; when $A<0, B<0$, because $x^*>1, or, x^*<0$, thus x^* is not the equilibrium point; at the same time $\left.\frac{\partial F(x)}{\partial x}\right|_{x=1} > 0, \left.\frac{\partial F(x)}{\partial x}\right|_{x=0} < 0$, thus $x=0$ is the only equilibrium point.

3.3. Research on the Control of Online Public Opinion in the Mode of Individual Help for Online Donation

Collections of participants' strategy:
The selection set of the government's strategy is{intervene, do not intervene}; The selection set of the parties'(including recipients and enterprises) strategy is{explain, do not explain}; The selection set of the netizens' strategy is {pursue,do not pursue}. Thus Table 5 shows the income risk matrix of the government-parties-netizens.

Strategy combination	government	Parties	Netizens
(intervene, explain, pursue)	q1-a	-b	M1-c+U
(intervene, explain, do not pursue)	q1-a+q2	θ1-b	U+M2
(intervene, do not explain, pursue)	q1-a-r1	-β1-β2	-w-m2-c+M1
(intervene, do not explain, do not pursue)	q1-a+q2-r1	θ2	-w+M2
(do not intervene, explain, pursue)	-r2	-b	-u1-m1-c+M1
(do not intervene, explain, do not pursue)	0	θ1-b	-u1+M2
(do not intervene, do not explain, pursue)	-r2	-β1	-u1-c-m1-m2-w+M1
(do not intervene, do not explain,do not pursue)	0	θ2	-u1-w+M2

Table 6 shows the related risk parameters involved in the income risk matrix.

Table 6.Parameter Definition

symbol	Symbol Description
U	In the intervention of government, there would be a stable income to the community when the parties choose to take the initiative to explain.
q1	Government can enhance the credibility from public due to intervention.
q2	Government can obtain the additional benefits due to public not pursuing in the intervention of government.
a	Government's intervention need to pay the cost.
r1	Governments are subject to public pressure in the negative response from the parties when they take the intervention.
r2	Government's non-intervention cause the public to account for the decline in its credibility
u1	Government's non-intervention hinder the development of Internet charity, damage the public's vital interests.
b	Explaination costs of the parties.
θ1	Parties receive additional benefits from public understandings by explaining .
β1	Parties are damaged their own image from responding negatively as well as accepted social condemnation and public opinion.
β2	In the intervention of the government, parties would accept more serious impact, moral and even legal punishment without explaining.
w	Parties bring the serious loss to the public without explaining.
θ2	Parties obtain the additional benefits from public understanding without explaining.
c	Public's need to pay the cost with pursuing .
M1	Public can obtain the inherent gains with pursuing.
M2	Public can obtain the inherent gains without pursuing .
m1	Public receive the psychological gap due to government's non-intervention.
m2	Public receive the psychological gap due to parties' non-explaination.

Assuming that the proportion of government chooses to active intervene is x, the ratio of non-intervention is (1-x); the ratio of the parties to choose to explain is y, then the ratio is not explained (1-y); the ratio of public choice is Z, the ratio is not (1-z);
The government chooses the benefits of different strategies available as:

$$A_{11} = yz(q_1 - a) + y(1-z)(q_1 - a + q_2) + (1-y)z(q_1 - a - r_1) + (1-y)(1-z)(q_1 - a + q_2 - r_1)$$
$$A_{12} = yz(-r_2) + (1-y)z(-r_2) = -zr_2$$
$$\overline{A_1} = xA_{11} + (1-x)A_{12}$$

The government chooses to intervene with the dynamic replication equation can be referred as follows:

$$F(x) = \frac{dx}{dt} = x(A_{11} - \overline{A_1}) = x(1-x)(yq_1 - ya + yq_2 + q_1 - a + q_2 - r_1 - zq_1 + za + zr_1 + zr_2) \quad (6)$$

Make $F(x) = 0$,so that the balance of the tripartite game is $z = \frac{-r_1 + q_1 - a + q_2}{q_1 - a - r_1 - r_2} + \frac{q_1 - a + q_2}{q_1 - a - r_1 - r_2} y, x = 1,0$, by virtue of the last method,which means the government will choose to take the initiative intervention or not to intervene.
Similarly, the benefits of the different strategies are chosen by the parties as follows:

$$B_{11} = xz(-b) + x(1-z)(\theta_1 - b) + (1-x)z(-b) + (1-x)(1-z)(\theta_1 - b)$$
$$B_{12} = xz(-\beta_1 - \beta_2) + x(1-z)\theta_2 + (1-x)z(-\beta_1) + (1-x)(1-z)\theta_2$$
$$\overline{B_1} = yB_{11} + (1-y)B_{12}$$

Then the replication dynamic equation of the parties interpreting :

$$F(y) = \frac{dy}{dt} = y(B_{11} - \overline{B_1}) = y(1-y)(\theta_1 - b - z\theta_1 + xz\beta_2 + z\beta_1 - \theta_2 + z\theta_2) \quad (7)$$

Make $F(y) = 0$ so that the balance point of the three-way game is $x = \frac{\theta_2 - \theta_1 + b}{\beta_2}\frac{1}{z} + \frac{\theta_1 - \theta_2 - \beta_1}{\beta_2}, y = 1,0$, by virtue of the last method, which means Parties will be choose to explain and confess the event or not to explain the incident.
Another, the benefits of the different strategies are chosen by the netizens as follows:

$$C_{11} = xy(M_1 - c + U) + x(1-y)(M_1 - c - w - m_2) + (1-x)y(-u_1 - c - m_1 + M_1) + (1-x)(1-y)(-u_1 - c - m_1 - m_2 - w + M_1)$$
$$C_{12} = xy(U + M_2) + x(1-y)(-w + M_2) + (1-x)y(-u_1 + M_2) + (1-x)(1-y)(-u_1 - w + M_2)$$
$$\overline{C_1} = zC_{11} + (1-z)C_{12}$$

Then netizens choose to pursue with the dynamic replication equation can be referred as follows:

$$F(z) = \frac{dz}{dt} = z(C_{11} - \overline{C_1}) = z(1-z)(-m_1 - c + M_1 + xm_1 - m_2 + ym_2 - M_2) \quad (8)$$

Make $F(z) = 0$ so the balance of the tripartite game can be obtained $z = 1,0, y = \frac{m_1 + c + m_2 + M_2 - M_1}{m_2} - x\frac{m_1}{m_2}$, by virtue of the last method,which means netizens will choose to pursue relevant responsible party or not to pursue.

Government-parties-netizens' tripartite dynamic game corresponding to Jacobian matrix in the control phase of public opinion :

$$J=\begin{pmatrix} (1\text{-}2x)(yq_1-ya+yq_2+q_1-a+ \\ q_2-r_1-zq_1+za+zr_1+zr_2) & (x-x^2)(q_1-a+q_2) & (x-x^2)(-q_1+a+r_1+r_2) \\ (y-y^2)z\beta_2 & (1-2y)(\theta_1-b-z\theta_1+xz\beta_2+ \\ z\beta_1-\theta_2+z\theta_2) & (y-y^2)(-\theta_1+\theta_2+x\beta_2+\beta_1) \\ (z-z^2)m_1 & (z-z^2)m_2 & (1-2z)(xm_1+M_1-M_2-m_1 \\ -c-m_2+ym_2) \end{pmatrix}$$

Make $A=yq_1-ya+yq_2+q_1-a+q_2-r_1-zq_1+za+zr_1+zr_2$ $B=\theta_1-b-z\theta_1+xz\beta_2+z\beta_1-\theta_2+z\theta_2$ $C=xm_1+M_1-M_2-m_1-c-m_2+ym_2$

The determinant detJ and the trace TrJ of the Jacobian matrix are :

$$\begin{aligned} &TrJ=(1\text{-}2x)A+(1-2y)B+(1-2z)C \\ &\det J=(1-2x)(1-2y)(1-2z)ABC+xyz(1-x)(1-y)(1-z)[(q_1-a+q_2)(-\theta_1+x\beta_2+\beta_1+\theta_2)m_1 \\ &+(-q_1+a+r_1+r_2)m_2z\beta_2]-xy(1-x)(1-y)(1-2z)C(q_1-a+q_2)\beta_2-yz(1-y)(1-z) \\ &(1-2x)A(-\theta_1+x\beta_2+\beta_1+\theta_2)m_2-xz(1-x)(1-z)(1-2y)B(-q_1+a+r_1+r_2)m_1 \end{aligned} \quad (9)$$

According to the balance point analysis results, when the parties choose to explain the strategy, regardless of whether the netizens choose to pursue, because $q_1+q_2-a>0$; regardless of whether the government intervention, when the netizens choose to pursue the strategy $z=1$, and if $\beta_1-b+\beta_2<0$ $M_1-M_2-c>0$ then the netizens choose not to pursue the strategy $z=0$, we can immediately and substitute $\theta_1-b-\theta_2>0$ $M_1-M_2-c<0$ into the formula(9), at the same time we can find that when the equilibrium point is $(0,0,0),(0,1,0),(1,0,0),(1,0,1)$, because $\det J<0$, the determinant is the saddle point and is in the local steady state, which means that the strategy is not enough to make the formation stage of public opinion continue to advance; when the equilibrium point is $(0,0,1),(0,1,1),(1,1,1)$, because $\det J>0$, TrJ uncertainty . Then the equilibrium point is still in the local stability state of the saddle point. When the equilibrium point is $(1,1,0)$ satisfied $\det J>0, TrJ<0$, the three parties' interests are equalized and the steady state is satisfied, then the equilibrium point is $(1,1,0)$ the evolution stabilization point. Indicating that the government chooses to take the initiative to intervene, parties take the explanation and netizens choose not to pursue the strategy for the evolution of stability strategy.

4. CONCLUSION AND PROSPECT

Based on the perspective of individual help on online fundraising model, taking the Rohr event as an example, the dynamic game model is used to analyze the different behavior strategies of participants in the evolution process of online public opinion under the influence of the event. The research shows that the multi-party participants joint in promoting the development of public opinion, «high interaction, low cost» characteristics of the social network platform and the «lack of supervision, the audit is not strict» drawbacks provide convenience and channels for the recipients releasing the fititious information , In the enterprise's cooperation, this kind of help information with the purpose of marketing will be forwarded and spread in netizens, promote the rapid formation of online public opinion, the exposure of the inside, the questioning and interaction between netizens promote the online public opinion to the next stage of evolution, this time the government and the relevant responsible party will take interpretation to control public relations fluctuations and negative effects, so as to meet the public interest and to restore public confidence in online public welfare. Therefore, this paper argues that, as an important channel for help, the relevant platform should establish a reasonable risk control mechanism, the personal helpers conduct a rigorous review of information on individual help ,to strengthen the authenticity of information, integrity of the screening, and real-time tracking the use of funds. From the source to prevent the occurrence of such incidents. At the same time, the government and charitable organizations should pay attention to individual help and other online public welfare activities in real time, implement social supervision, once the project found fraud, should immediately stop fundraising behavior to achieve donation earmarked and maximize the benefits. For netizens, they should also be rational for providing help, learn to identify the authenticity of help information, pay attention to the implementation of donations to ensure that their kindness to use maximum. In addition, the next step of the article will build the simulation analysis model by digging the case data to explore the evolution process of the online public opinion which affects the different strategies of the subject's behavior in the individual seeking network fundraising mode, so as to provide more targeted and guidance for the construction of the healthy online public ecological.

REFERENCES

- FANG Fu-jian (2011). Study on the Evolution of Public Opinion on Network of Unexpected Event. Wuhan: Huazhong University of Science and Technology.
- Shibo (2010). A Study of the Intrinsic Evolution Mechanism for Network Sentiment of Public Crisis Events. Journal of Intelligence,29,41-45.
- Zeng Runxi (2009). Research on the mechanism of network public opinion control. Library and Information Work,28,79-82.
- Bin Ning,Yank Shubiao and Hu Feng (2017). Game Analysis on Evolution of Internet Public Opinion of Emergency: Case Study of Maoming PX Incident. Information Research,01,32-39.
- Sznajd J N K. (2000). Opinion evolution in closed community.Inter-national Journal of Modern Physics C,11,1157-1165.
- Weisbuch G (2004).Bounded confidence and social networks. The Europe Physical Journal B-Condensed Matter and Complex Systems,38,339-343.
- WANG Shu-xian (2016). Tripartite Game Analysis of Public Opinion Evolution of Microblogging with Leaders' Participation.Researches In Library Science,01,19-25.
- SUN Yi-xiao (2017). Game Analysis of Online Public Opinion and Judicial Action.Legal Economy,4,120-124.
- WANG Xu,SUN Rui-ying (2017). Research about Online Public Opinion Spread during Emergencies Based on Social Network Analysis——A Case Study of thewei Zexi. Information Science,3,87-92.
- QI Kai, YANG Zhi and ZHANG Zi-mo (2017). Analysis on the Evolutionary Game of Public Opinion Guidance Mechanism ofnetizens under the Participation of Government. Information Science,3,47-52.
- ZHANG Li-fan,CHENG Nan and ZHU Heng-min (2017). Analysis of Network Public Opinion Mechanism Based on Dynamic Game Theory and Media Participation .Information Science,1,144-147.

Author Text

Zhang Lin, Department of Management Information System,Tianjin University of Finance and

Economics,China,276963955@qq.com

The recovery approach to the silent outbreak of Mental Illness in People of African descent.

Charmaine Tandiwe Kuwana,
Master in Mental Health Nursing, Central Queensland University, Australia

ABSTRACT

The rate of suicide and mental illness is on the increases in people from Culturally and Linguistically Diverse communities (CaLD). Is it imperative to discuss the rationale behind the lack of engagement with mental health services from people from CaLD backgrounds. This paper claims to offer an alternative to the current practice that is holistic in nature and addresses the spiritual needs of people from Africa. The evidence used in this article was from studies conducted on cultural beliefs surrounding mental illness and lack of understanding of mental illness by migrants coming from developed countries. Current evidence suggests that there is need for future research to address mental health and its variants as understood by people from the developing countries. As societies from developing countries transition from traditional ways to modern lifestyles, the issues many women and families are facing is the cultural struggle between women's traditional role and their modern traditional role.

INTRODUCTION

The author works in a Mental Health community team for adults aged from 18 to 65 years old. The teams cater for about 800 clients and the majority of them are Caucasians from various backgrounds. Of the 50 members only 2 are African. These comprise of a social worker and a Registered Mental Health nurse, who happens to be the author. The target audience for the Recovery education resource is members of staff including management. Although Africa is a is a continent, most clinicians and lay people tend to view this diverse population as a homogenous society. However, most of the cultures and subcultures share some similarities. The author wanted to explore the meaning of recovery from her own Zimbabwean subculture and from a mental health perspective. Having had the lived experience of immigrating from a developing county to United Kingdom then Australia, the author still feels there is not enough known in mental health practice about immigrating coupled with mental illness and recovery from the eyes of the immigrant. Evidence of the importance of lived experience of mental health challenges is discussed by (Happell et al., 2015) and the author wishes to dispel misconceptions with regards to the culture of people from African countries who access services.

The author thought it would be a good opportunity to discuss cultural competency and educate her workmates about her culture and how she has embraced the recovery model without compromising on her beliefs. Race, culture and ethnicity can be sensitive issues to talk about so she wanted colleagues to feel free to ask her questions they otherwise would not be comfortable asking ethnic minorities mental health consumers. The author opted for an oral presentation in a casual format

There has been literature and journals written about ethnic minorities but it does not cover the core beliefs they hold about mental health in African countries and their religion. She thought she could create a dialogue within her colleague about her beliefs and how this impacts on medication concordance and recovery. In Western Australia the number of ethnic minorities accessing services is still small but it's on the increase and by empowering clinicians with the knowledge about religion and culture they might find it easier to raise the subjects with mental health consumers. They will be more likely to support consumers on their recovery journey if they understand what illness, cause and recovery mean to them.

(Drummond, Mizan, Brocx, & Wright, 2011) discusses the stigma linked to certain illness within the African cultures, suspicions about the Western medications and how this can hinder immigrants from engaging with mental health services. The author grew in a society that viewed mental challenges as only experienced by people of low socio-economic status and was caused by witchcraft. It's through studying mental health nursing that she shifted her way of thinking but from speaking from fellow countrymen such beliefs are still highly held. That being said, to be cured of such mental challenges one has to go Church and have church elders pray for them and these beliefs are still followed in many African immigrant households in Australia. Religion and spirituality are of paramount importance in our culture.

Most believe when one has gone through mental challenges and they get better God has intervened. This is a challenge to clinicians who might otherwise be unaware of such practices as once God has intervened there is no need to engage in services that are shunned by the community or take western medications. This could lead to consumers not engaging or being avoidant of clinicians and lying about taking medications. This however could be likened to understanding recovery in the context of 'recovering from' that (De Vecchi, Kenny, & Kidd, 2015) explains as a biomedical classification of treating illnesses or getting rid of symptoms.

(Carlisle, 2015) realised from her research that clinicians were viewed as not interested in assisting consumers meet their spiritual needs or within their professional scope of practice. (Harris et al., 2015) states that despite the role of spirituality and religion some clinicians are not comfortable discussing the above with clients in fear that it would strengthen religious delusions in consumers. While there is some truth to this (Milner & Khawaja, 2010) advocates for clinicians to have a strong understanding of African Refugee's background, dilemmas and psychosocial stressors. Most of the African immigrants' experience acculturation stressors and these include disintegration of social and cultural norms and loss of social support from extended families. (Ho et al., 2016) concluded that spiritual experiences may be seen as signs and symptoms of mental challenges and this kept consumers from discussing religion with clinicians thereby neglecting a potential positive coping mechanism. (Harris et al., 2010) discovered that amongst mental health consumers, the more religious ones describe having a better quality of life, positive outcomes and better adaptive coping skills. The author wanted to equip her team members with coping resources they could use when engaging with some African consumers who subscribe to the same values and beliefs. The above highlight why the author chose to speak about culture, religion and spirituality and how understanding some of these confusing aspects can help them in dealing with consumers from different backgrounds as them. With the management present during the presentation the author hoped they would talk to stakeholders about liaising with African community groups about offering support to some African immigrants who are deemed as needing support integrating into the Australian culture whilst staying true to theirs

THE NEED FOR MENTAL HEALTH CULTURAL COMPETENCY

Having come from a country that believes that mental illness is caused by witchcraft the author has had to do practice a lot self-awareness. (Jack, 2008) describes self-awareness as the ability to understand our own emotions, strengths and weaknesses and the impact of our their behaviours in adverse scenarios. The author imagines how some discussions about religion might be anxiety provoking for clinicians who do not share the same sentiments about religion or have their own prejudices. The author is a non-practicing religious person and when she goes through rough patches in her life she calls upon God for help and believes that God answers her prayers. The author believes her faith aids her recovery journey as described by (Harris et al., 2015) This is a very difficult concept to understand for some clinicians who are not religious. Elouo et al (2009) explains how some communities around the world attribute the cause of mental illness to misfortune, supernatural causes and witchcraft. This belief contributes to the stigma of mental health in ethnic minorities. Depression is viewed as a manifestation of possession of evil spirits that has entered their body and is controlling them. An understanding of these beliefs will help health departments plan and implement how they are how to deliver care in a culturally sensitive manner.

I have gone through some trying times and could have been depressed at some point but in my culture there is no word for depression in my language so it was difficult to get help from people from the same background. (Benson, Haris, & Saaid, 2010) address what the author is trying to explain that most African languages do not have words for what clinicians use to describe mental health symptoms. They go on to explain how these symptoms may be described as 'high blood pressure or BP'. This is easily understood across many African countries as very stressful situations, trying times and a crisis. Different mental health illnesses are not endorsed and are often interpreted in idiom of distress. The expression of emotion may be viewed as shameful hence mood disturbance is not verbalized. Lawati et al, (2002) concluded that western cultures have a tendency to 'psychologise' distress which is in contrast to non-western culture where distress is 'somatised.'

I imagined myself when I first immigrated and the stress it brought having to adapt to another culture whilst staying true to myself. That can be very hard for an ethnic minority who has a mental illness. The stigma attached to being admitted in a mental health hospital is twice-fold compared to that Westerners. The community may shun you and you become the subject of talk. The religious belief that when unwell God intervenes and heals you is still my source of strength. The thought of the community knowing that one is accessing mental health services is daunting.

My experience of working in Mental Health Services taught me that there is place for medications and recovery is a journey not being cured. The consumer will need to feel supported in meeting their spiritual needs despite what the clinicians believe in. There is still a gap in practice about meeting the spiritual needs of consumers especially when they are psychotic. Clinicians will have to be careful in assessing whether their religions are not part of their symptomology and entwined in their delusions. Some of the religious beliefs held in Africa or other parts of the worlds are symptoms of mental health conditions. Family assessment and involvement might be useful in explaining cultural beliefs and traditions. Clinicians will have to be non-judgmental as most people from the author's culture would not openly discuss their religious beliefs and traditions if they are sometimes classified as symptoms of mental illness. For example, 'some might believe that God speaks to them and they have conversations with God but this could be an auditory hallucination or delusion in Mental health.

THE ROLE OF MENTAL HEALTH SERVICES IN ADVOCATION FOR BETTER UPTAKE OF RECOVERY PRINCIPLES

This part of the paper, the author uses an acronym (SPICES) Spiritual, Physical, Intellectual, Cultural, Emotional and Social in order to cover all aspects of a being and be holistic. Each aspect of the acronym is explored with regards to how mental health services can advocate for recovery uptake in African immigrants who are also mental health consumers.

Spiritual

As discussed above spirituality plays a huge part in the lives of many African immigrants and is a deep source of strength, courage, wisdom when they are faced with difficult situations. (Mahmood, Moore, Savic, & Chur-Hansen, 2016) highlights that in the Sudanese culture people do not go to see a special person for emotional distress, grief and as these are seen as part of life so there is resistance within the community towards medicalisation of emotional distress. The clinician's role could be supporting consumers in doing what works for them and this could be organizing with family to talk to church elders who are sometimes involved in praying for the consumer. With the consent of the consumer and their family the service could establish partnerships with church elders from the same ethnic background to visit clients and pray with them. The nurses could support the consumer by letting the consumer teach clinicians about their faith and significant support systems. (Slade, 2011) explains how nurses can break boundaries and how this can help the relationship between nurse and consumer by something as simple as having clinicians talk about their own spiritual beliefs and show a level of humility.

Physical

The role of the nurse is to assist the consumer in weighing their options about health behaviours such as medications management and help them project foreseeable outcomes of different choices as advised by (Carlisle, 2015). The consumer from an ethnic minority background might decide to stop taking their medications because they believe God had healed them. The clinicians could support the consumer in making that decision and enable them to take positive risks. Psycho-education would be paramount in situations like this as the consumer would have to be well educated on their mental health condition, the benefits of taking their medications and the likely result of not including support systems for when they feel they are deteriorating. Family could be educated on the mental health condition and symptoms to watch out for. The clinician would enable the consumer make informed decisions that affect their recovery journey.

Intellectually

Most consumers still believe clinicians are experts at mental health illnesses instead of their lived experiences as discovered by (Topor et al., 2006) that consumers had greater confidence in what clinicians said as opposed to lay people. Most immigrants from Africa come from cultures that do not recognize mental health symptoms and do not have word for them. Consumers look up to clinicians for knowledge of different mental health challenges they might have experienced so nurses could play the role of an educator. By educating the consumer about mental health illnesses, symptoms, triggers, medication, and services available. This will not only empower the consumer but equip them to become knowledgeable partners in care.

Cultural

Culture is a way of life and helps people make sense of surroundings, behavior and values. (Hammond 1978) advocated for clinicians to learn about consumer's belief systems in order to show respect and also acknowledge that the consumer is the expert in their own culture. It shows that the clinician is not dealing with the consumer from their own ethnocentric beliefs. This calls for clinicians to be creative in their approach such as using interventions that may exist outside their service/team and these could come from the consumer as suggested by (Keeling & McQuarrie, 2014). An example that comes to mind could be in some cultures witch-doctors/ give sick people water mixed Epsom salts and herbs to spray in their rooms to get rid of evil spirits. Such activities could be aimed at meeting specific needs of a consumer only understood by the consumer

Emotional

The emotional aspects of a being could be regarded as how they view themselves and their worth. Most immigrants already struggle with issues of racism and see themselves as inferior to their white counterparts possibly due to colonization. Giving them a purpose could aid in their journey to recovery. (Trossman, 2011) speaks highly of a service that offers 3 month contracts to consumers and those who do well are given jobs in hospitals. She speaks of how mental health nurses can support those who are capable of working access work in not so stressful environments. Nurses could champion this cause and make referrals to such organisations and be there as consumers' transition from not working to having to get up and go to work. One of the roles of the mental health nurses in recovery is supporting consumers set goals and assist them in breaking down the tasks so they are manageable and (van Langen, Beentjes, van Gaal, Nijhuis-van der Sanden, & Goossens, 2016) discovered consumers appreciated having someone help them break down goals to manageable tasks such as the goal of having paid employment

Social

(Khawaja, McCarthy, Braddock, & Dunne, 2013) concluded that many clinicians do not have an understanding of migration and how it affects mental health of refugees. Africa culture tends to be family orientated including having extended family. Immigration has meant that some of these family circles are broken and some consumers might find themselves isolated. In some subcultures some groups might have community elders who may assist the consumer in socio cultural issues. With consent from consumers, clinicians could support them meet their social needs by reaching out to the elders who may also act as appointed persons as stated by (Gibson, 2015) as he summarised the new Mental Health Act 2014 which identifies that the next of kin might not necessarily be the chosen appointed person to be involved in the consumer's care. Relationships, family members and social support systems all have a part to play to play in supporting people on their recovery journey as insisted by (Tilley & Cowan, 2011) when discussing the common agreed essentials to promoting recovery. The nurses have a role to advocate for the extension of such services for their consumers.

CONCLUSION

This paper has found there is limited knowledge of the role of spirituality and religion in mental health especially in people from cultural and linguistically diverse populations.

The literature search on mental health in migrant communities yields a small result. There is need to conduct more research on mental illness in culturally and linguistically diverse populations. Refugees and asylum seekers are particularly more vulnerable to mental health but there is lack of specialized mental health care in Australia. This paper does not address the psychosocial factors that could be contributing to the increased rate of mental illness. Access to culturally tailored mental health service should be part of the mental health agenda in Australia. Further research is imperative to ascertain preventative measures that can be taken to promote mental health in culturally and linguistically diverse populations.

REFERENCES

- Benson, J., Haris, T. A., & Saaid, B. (2010). The meaning and the story: reflecting on a refugee's experiences of mental health services in Australia. *Mental Health In Family Medicine, 7*(1), 3-8.
- Carlisle, P. A. (2015). "We don't Talk about that Around Here": religion, Spirituality and Mental Health in Northern Ireland. *Mental Health, Religion & Culture, 18*(5), 396-407. doi:10.1080/13674676.2015.1071787
- De Vecchi, N., Kenny, A., & Kidd, S. (2015). Stakeholder views on a recovery-oriented psychiatric rehabilitation art therapy program in a rural Australian mental health service: a qualitative description. *International Journal of Mental Health Systems, 9*(1), 1-11. doi:10.1186/s13033-015-0005-y
- Drummond, P. D., Mizan, A., Brocx, K., & Wright, B. (2011). Barriers to Accessing Health Care Services for West African Refugee Women Living in Western Australia. *Health Care for Women International, 32*(3), 206-224. doi:10.1080/07399332.2010.529216
- Eloul, L, Al-Aldawi (2009) Silent epidemic of depression in women in the Middle East and North Africa Region.Emerging tribulation or fallacy? *Sultan Qaboos University Journal, 9*(1), 5-15
- Gibson, N. (2015). Mental health: Mental health act 2014: Doc, are you ready? Medicus, 55(09), 7.
- Hammond, Peter B (1978). *An introduction to cultural and social anthropology* (2d ed). Macmillan, New York
- Happell, B., Bennetts, W., Harris, S., Platania-Phung, C., Tohotoa, J., Byrne, L., & Wynaden, D. (2015). Lived experience in teaching mental health nursing: Issues of fear and power. *International Journal of Mental Health Nursing, 24*(1), 19-27. doi:10.1111/inm.12091
- Harris, J. I., Erbes, C. R., Engdahl, B. E., Tedeschi, R. G., Olson, R. H., Winskowski, A. M. M., & McMahill, J. (2010). Coping Functions of Prayer and Posttraumatic Growth. *International Journal for the Psychology of Religion, 20*(1), 26-38. doi:10.1080/10508610903418103
- Harris, J. I., Nienow, T., Choi, A., Engdahl, B., Nguyen, X. V., & Thuras, P. (2015). Client report of spirituality in recovery from serious mental illness. *Psychology of Religion and Spirituality, 7*(2), 142-149. doi:10.1037/a0038538
- Ho, R. T. H., Chan, C. K. P., Lo, P. H. Y., Ping Ho, W., Chan, C. L. W., Leung, P. P. Y., & Chen, E. Y. H. (2016). Understandings of spirituality and its role in illness recovery in persons with schizophrenia and mental-health professionals: a qualitative study. *BMC Psychiatry, 16*, 1-11. doi:10.1186/s12888-016-0796-7
- Jack, K. (2008). Exploring self-awareness in mental health practice. *Mental Health Practice, 12*(3), 31-35.
- Keeling, J. L., & McQuarrie, C. (2014). Promoting mental health and wellbeing in practice. *Mental Health Practice, 17*(5), 26.
- Khawaja, N. G., McCarthy, R., Braddock, V., & Dunne, M. (2013). Characteristics of culturally and linguistically diverse mental health clients. *Advances in Mental Health, 11*(2), 172-187. doi:10.5172/jamh.2013.11.2.172
- Mahmood, M. A., Moore, V. M., Savic, M., & Chur-Hansen, A. (2016). 'We don't have to go and see a special person to solve this problem': Trauma, mental health beliefs and processes for addressing 'mental health issues' among Sudanese refugees in Australia. *International Journal of Social Psychiatry, 62*(1), 76-83. doi:10.1177/0020764015595664
- Milner, K., & Khawaja, N. G. (2010). Sudanese Refugees in Australia: The Impact of Acculturation Stress. *Journal of Pacific Rim Psychology, 4*(1), 19-29. doi:10.1375/prp.4.1.19
- Slade, M. (2011). Breaking down boundaries. *Mental Health Practice, 15*(2), 9.
- Tilley, S., & Cowan, S. (2011). Recovery in mental health policy: good strategy or bad rhetoric? *Critical Public Health, 21*(1), 95.
- Topor, A., Borg, M., Mezzina, R., Sells, D., Marin, I., & Davidson, L. (2006). Others: The Role of Family, Friends, and Professionals in the Recovery Process. *American Journal of Psychiatric Rehabilitation, 9*(1), 17-37. doi:10.1080/15487760500339410
- Trossman, S. (2011). Overcoming stigma: Education and advocacy can make a difference in mental health care and services. *American Nurse, 43*(2), 1-9.
- van Langen, W. J. M., Beentjes, T. A. A., van Gaal, B. G. I., Nijhuis-van der Sanden, M. W. G., & Goossens, P. J. J. (2016). How the Illness Management and Recovery Program Enhanced Recovery of Persons With Schizophrenia and Other Psychotic Disorders: A Qualitative Study. *Archives of Psychiatric Nursing*, 30, 552-557. doi:10.1016/j.apnu.2016.04.005

The new development of qualitative research approach in language teaching and learning

Sainan Zhang,
Master, Tianjin University of Finance & Economics, China

ABSTRACT

From a language teaching and learning perspective, great developments have taken place in the field of qualitative research since the beginning of the millennium, which is not only a period of valuable consolidation but also with signs of new avenues opening up. According to John W. Creswell, the qualitative research has been mainly classified into five approaches. In the area of language teaching and learning area, the qualitative research methods used have a remarkable growth. And this article adopts three of them (narrative studies, ethnography and case study) to illustrate how to do a practical research in the qualitative research method with a typical corresponding sample study chosen from the TESOL Quarterly because of its applied linguistics and educational background context, as a result of which may give some inspirations for the development of qualitative research in the applied linguistics in the future.

1. QUALITATIVE RESEARCH

When it comes to qualitative research, Holliday (2007:1) notes that it is traditional to begin by distinguishing qualitative research from quantitative research. And for Frederick Erickson, qualitative inquiry seeks to discover and to describe narratively what particular people do in their everyday lives and what their actions mean to them. It identifies meaning-relevant kinds of things in the world—kinds of people, kinds of actions, kinds of beliefs and interests—focusing on differences in forms of things that make a difference for meaning.

'Qualitative research', claimed Lazaraton (2003a) in a review of evaluative criteria, 'has come of age in applied linguistics'. In the thirteen years since her own groundbreaking assessment enough has changed to warrant this claim. Qualitative research has opened dimensions of insight into the processes of language teaching and learning that were not even discernible on the horizon twenty years ago, and developments in the new millennium promise even richer understandings in the future. And in the field of applied linguistics, the section of language learning and teaching, is using increasingly sophisticated methodological approaches to address a growing number of urgent real-word problems. Many researchers have done a lot of research in this field, they have carefully talked about the trends and the publication of qualitative articles in some learning and teaching journals.

2. SAMPLE COLLECTION

Language teaching and learning journals appear to be far more open to the publication of qualitative data-based research than they were in the early 1990s. (Phil Benson 2009) after a decade in which reliance on qualitative methods were questioned, Magman (2005) argued, "our discipline now embraces a variety of qualitative methods as accepted or even preferred, methods of inquiry." Published surveys of journal articles lend some support to this view, like Language Learning, Studies in Second Language Acquisition and TESOL Quarterly.

For this study, I've chosen TESOL Quarterly as a research sample. TESOL Quarterly is a quarterly peer-reviewed academic journal published on behalf of TESOL International Association. It covers English language teaching and learning and standard English as a second dialect, including articles on the psychology and sociology of language learning and teaching, professional preparation, curriculum development, and testing and evaluation. According to the Journal Citation Reports, the journal has a 2013 impact factor of 1.000, ranking it 41st out of 169 journals in the category «Linguistics» and 64th out of 219 journals in the category «Education & Educational Research». Considering the research question of this article, TESOL Quarterly is a very good example that combines the language teaching and learning, linguistics and the research methodology.

In recent years, qualitative research has experienced great development across the worldwide. Qualitative research has been widely used in applied linguistics and it has won fruitful research results. As Lazaraton (2003a) claimed in a review 'Qualitative research has come of age in applied linguistics.' And Benson et al also think (2009) that Language teaching and learning journals also appears to be far more open to the publication of qualitative data-based studies than they were in the early 1990s. Published sur veys of journal articles lend some support to the view. Surveying the contents of Language Learning, MLJ, Studies in Second Language Acquisition, and TESOL Quarterly from 1991 to 1997, Lazaraton (2002) found that 10% of the data-based articles used qualitative approaches. Extending the survey up to 2001, Lazaraton (2005) found that this figure had risen to 14%, with the proportion of qualitative articles reaching 40% in TQ. If you still have doubts about this statistics, Richards (2009) also put TESOL Quarterly into the "more than 10% qualitative research" journals' camping. These results explicitly explain the reason why I will choose TESOL Quarterly as the research sample for this paper, because it is always a great player in the stage of qualitative studies in the language teaching and learning area, even we could call it a 'hero' of qualitative research in the applied linguistics.

3. THEORETICAL BASIS

According to the book written by John W. Creswell and published in 2013 named QUALITATIVE INQUIRY & RESEARCH DESIGN: Choosing Among Five Approaches: He introduced five different qualitative study approaches, which is the theoretical base of this paper. And they are as follows: narrative research; phenomenological research; grounded theory research; ethnography research; case studies.

Narrative research has many forms, uses a variety of analytic practices, and is rooted in different social and humanities disciplines. A phenomenological study describes the common meaning for several individuals of their lived experiences of a concept or a phenomenon. The grounded theory is a qualitative research design in which the inquirer generates a general explanation (a theory) of a process, an action, or an interaction shaped by the views of a large number of participants. An ethnography focuses on an entire culture-sharing group in which the researcher describes and interprets the shared and learned patterns of values, behaviors, beliefs, and language of a culture-sharing group. Case study is a type of design in qualitative research that may be an object of study, as well as a product of the inquiry. Case study research is a qualitative approach in which the investigator explores a real life, contemporary bounded system (a case) or multiple bounded systems (cases) over time, through detailed, in-depth data collection involving multiple sources of information (e.g., observations, interviews, audiovisual material, and documents and reports), and reports a case description and case themes. Table 1 shows the contrast characteristics of these five qualitative approaches:

Table 1. Contrasting Characteristics of Five Qualitative Approaches

Characteristics	Narrative Research	Phenomenology	Grounded Theory	Ethnography	Case Study
Focus	Exploring the life of an individual	Understanding the essence of the experience	Developing a theory grounded in data from thefield	Describing and interpreting a culture-sharing group	Developing an in-depth description and analysis of a case or multiple cases
Type of Problem Best Suited for Design	Needing to tell stories of individual experiences	Needing to describe the essence of a lived phenomenon	Grouding a theory in the views of participants	Describing and interpreting the shared patterns of culture of a group	Providing an in-depth understanding of a case or cases
Discipline Background	Drawing from the humanities including anthropology, literature,history,psychology,and sociology	Drawing from philosology,psychology,and education	Drawing from sociology	Drawing from anthropology and sociology	Drawing from psychology,law,political science,and medicine
Unit of Analisis	Studying one or more individuals	Studying several individuals who have shared the experience	Studying a process ,an action,or an interaction involving many individuals	Studying a group that shares the same culture	Studying an event,a program,an activity,or more than one individual

Characteristics	Narrative Research	Phenomenology	Grounded Theory	Ethnography	Case Study
Data Collection Forms	Using primarily interviews and documents	Using primarily interviews with individuals,although documents,observations,and art may also be considered	Using primarily interviews with 20-60 individuals	Using primarily observations and interviews,but perhaps collecting other sources during extended time in field	Using multiple sources,such as interviews,observations,documents,anc artifacts
Data Analysis Strategies	Analyzing data for stories,"restorying" stories,and developing themes,often using a chronology	Analyzing data for significant statements,meaning units,textual and structural description,and description of the "essence"	Analyzing data through open coding,axial coding,and selective coding	Analyzing data through description of the culture-sharing group and themes about the group	Analyzing data throuth description of the case and themes of the case as well as cross-case themes
Written Report	Developing a narrative about the stories of an individual's life	Describing the "essence" of the experience	Generating a theory illustrated in a figure	Describing how a culture-sharing group works	Developing a detailed analysis of one or more cases

After careful reading through the articles of TESOL Quarterly in the recent years, I have found that the five qualitative research methods don't have an equal role to play in this journal. What's more, I am not alone to think in this way. Other researchers also hold the same opinion as I do, However, while phenomenology and grounded theory feature in qualitative approaches generally, they do not loom large in research in language teaching and introspective methods are as likely to be quantitative as qualitative(Richard 2009)the percentage of phenomenological study and the grounded theory research is very low in the published articles. Considering the definitions of these five approaches, these two approaches are quite difficult to put into practice.

So in this article, I have chosen the narrative studies, the ethnography research and also the case study as research focus. For each qualitative research approach, I have chosen a sample article to illustrate how this kind of approach is used in the practical study in the section of language teaching and learning section.

4. THREE SAMPLE STUDIES

4.1 A Narrative Study

A.Suresh Canagarajah (2012) is a good example of a typical narrative study in language learning and teaching in TESOL Quarterly. The title of this article is Teacher Development in a Global Profession: An Autoethnography, which has clearly pointed out that the research methodology this article chooses. This article was published in TESOL Quarterly in 2012. According to Creswell's classification of the types of qualitative study, this article belongs to narrative study. In this article, the author represents the ways in which he negotiated the differing teaching practices and professional cultures of the periphery and the center in an effort to develop a strategic professional identity. Through his journey of professionalization, he talks about his own experiences like some vivid and lived stories of being an English teacher in Sir Lanka from1984 to 2012, when he has become a professor in Pennsylvania State University. And narrative also plays a very important role in the procedures of analyzing the stories told.

In order to conduct a narrative study, the author has followed the procedures bellow:
Determine if the research problem or question best fits narrative research. Narrative research is best for capturing the detailed stories or life experiences of a single individual or the lives of a small number of individuals. In this ethnographic self-reconstruction, the author represents the ways in which he negotiated the differing teaching practices and professional cultures of the periphery and the center in an effort to develop a strategic professional identity. Based on the author's experience, he developed a perspective on more egalitarian and dialogical relationships be

tween global teaching communities in TESOL and advocate for professional identities of English teaching.

Select one or more individuals who have stories or life experiences to tell, and spend considerable time with them gathering their stories through multiples types of information. Considering the author's professional experiences are very complex, the author chose himself as the research participants. He wrote his stories in journals and diaries. And for this article, the journals and diaries become a very important part of data analysis.

Consider how the collection of the data and their recording can take different shapes. In this article, writing is not only the means of disseminating one's knowledge and experiences. And the author admitted that "there is an emphasis on the creative sources of writing, especially narrative, for generating, recording, and analyzing data. Many forms of written artifacts go into the construction of my narrative: books and articles I read, institutional reports and correspondence about my professional performance." And these different data collection, which as the author hoped "generate additional comparisons and interpretations from alternate perspectives."

Collect information about the context of these stories. Narrative researchers situate individual stories within participants' personal experiences (their jobs, their homes), their culture (racial or ethnic), and their historical contexts (time and place). The author said in the conclusion of this article: "Perhaps it is the boundary-crossing work of transient, migrant, and refugee teachers like me that has led to the dominant paradigms being reconstructed in our profession." At the beginning, the author worked as an English teacher in Sir Lanka in 1984. Recognizing his limitations in the professional development, he went to America to further his studies of an educational doctor. Then he went back to his homeland. Unfortunately, due to the civil war at that time, he had to run away from his motherland as a refugee. And he came back to the USA, he works as a professor and teaches in the departments of English and Applied Linguistics at Penn State. So this research article includes all the elements aforementioned.

Analyze the participants' stories. The researcher may take an active role and "restory" the stories into a framework that makes sense. Restorying is the process of reorganizing the stories into some general type of framework. This framework may consist of gathering stories, analyzing them for key elements of the story (e.g. time, place, plot, and scene), and then rewriting the stories to place them within a chronological sequence.

Collaborate with participants by actively involving them in the research. The form of research is conducted and represented from the point view of the self. So the participant is definitely involved in this research.

Given these procedures and the characteristics of narrative research, narrative research is a challenging approach to use. The researcher needs to collect extensive information about the participant, and needs to have a clear understanding of the context of the individual's life. It takes a keen eye to identify in the source material that gathers the particular stories to capture the individual's experiences. Multiple issues arise in the collecting, analyzing, and telling of individual stories. Apparently, in this autoethnography research, the methodology is quite different from the traditional positivist research traditions that perceive anything based on the self as subjective and distorting valid knowledge claims, autoethnography values the self as a rich repository of experiences and perspectives. And in order to make the analysis explicit, the author also adopts an analytical autoethnography. And finally we can conclude that one can articulate one's own experiences, rather than letting others represent them. This is especially important for members of marginalized communities who lack the resources and publishing outlets to articulate their knowledge and interests. Their knowledge is often presented by outsiders according to outsiders' perspectives. Take this article for example, autoethnography is a valuable form of knowledge construction in the language teaching and learning field.

4.2. An Ethnographic Study

For the ethnography study, I have chosen an article named Addressing racialized multicultural discourses in an EAP textbook: working toward a critical pedagogies approach. The article was published in TESOL Quarterly in March, 2016 and it is a typical example of how ethnography research method is used in applied linguistics area. The author first presents the chapter's passages and then describes how the discourses were initially taken up by the participants. A conversation between the instructor and the author in which the instructor shared a racializing experience is then featured providing the context for her subsequent approach in which she revisited the reading with students. The author examines the classroom interactions with two questions in mind: " How did the instructor's critical pedagogies approach work to mediate the racialized representations in the chapter? Did this approach facilitate more meaningful dialogical engagements enabling the students to develop their academic literacies?"

In order to conduct an ethnography study, the author has followed the procedures bellow:
Ethnography focus on developing a complex, complete description of the culture of a group, a culture-sharing group. If a researcher wants to do an ethnography research, he or she should first determine if ethnography is the most appropriate design to use to study the research problem and identify and locate a culture-sharing group to study. For this article, the racialized multicultural discourses emerge in the TESOL classroom via textbook representations of immigrant success stories and perceived racial and cultural differences among students. Although liberal multicultural discourses may be well intentioned, these discourses warrant closer examination for the ways in which they can essentialize cultural identities and enact power dynamics of who is defining and who is being defined. Drawing on an ethnographic English for academic purposes (EAP) classroom case study, and with the aim of bridging the gap between critical theories and actual classroom practices, this article explores the approaches an instructor implemented with students in addressing such discourses in an EAP textbook chapter. The researcher looks for patterns of social organization (social networks) and ideational systems (worldview, ideas) (Wolcott, 2008a). This research was conducted in an EAP classroom, in an intensive English program at a Canadian public university. This means that the culture-sharing group has been intact and interacting for long enough to develop discernible working patterns. The ethnographic research involved observing the instructor's classes for 11 months (covering the four terms) and included audio- and video-recording her classes, interviews with her and the students, numerous research discussion meetings with the instructor (totaling 18 during the study), classroom observation field notes, and curriculum material analysis. The data presented here are drawn from two classroom lessons during this summer term class and from one of the meetings with the instructor. In addition, theory plays an important role in focusing the researcher's attention when conducting an ethnography. Using the theory and looking for patterns of a culture-sharing group involves engaging in extensive fieldwork, collecting data primarily through interviews, observations, symbols, artifacts, and many diverse sources of data. For example, ethnographers start with a theory—a broad explanation as to what they hope to find—drawn from cognitive science to understand ideas and beliefs, or from materialist theories. When it comes to the theory issues, the author talked about the liberal and critical multiculturalism, and also the critical pedagogies approach in language learning and teaching.

In an analysis of this data, the researcher relies on the participants' views as an insider emic perspective and reports them in verbatim quotes, and then synthesizes the data filtering it through the researchers' etic scientific perspective to develop an overall cultural interpretation. Inasmuch as social actors take up discourses differently partly based on their inscribed personal experiences, or their "historical bodies", the analysis of the instructor's teaching approaches toward the reading during the two class lessons takes into account her own inscribed experiences—both pedagogical and personal—that she shared in the research meeting featured below, which was part of the overall analysis. And in analyzing the classroom interactions, and the author's role as a researcher engaged in mediated collaborative actions that influenced ensuing teaching approaches, the author has employed a discourse analysis approach which holds that "the relationship of text to text, language to language, is not a direct relationship but is always mediated by the actions of social actors". The author's reading of the chapter's cultural representations and the participants' ensuing classroom discourses should also be seen as stemming in part from his lived experiences frequently encountering racialized discourses. This analysis results in an understanding of how the culture-sharing group works, the essence of how it functions, the group's way of life. The analytical aim of this article is to examine these approaches in the ways in which they engaged with the racialized discourses and the resulting pathways of the participants' meaning-making in their particular discursive actions.

4.3. A Case Study

For the case study approach, I have chosen an article named Imagined Community Falling Apart: A Case Study on the Transformation of Professional Identities of Novice ESOL Teachers in China The article was published in TESOL Quarterly in 2012 and it is a typical example of case study applied in the area of language teaching and learning. This qualitative case study described a case study on the transformation of professional identities of novice ESOL teachers in China.

In order to conduct an ethnography study, the author has followed the procedures bellow:
Case study research begins with the identification of a specific case. The intent of conducting the case study is also important. As many novice teachers experience a reality shock during the transition from teacher education programs to the first years of teaching considering the unpredictable and dynamic nature of authentic educational contexts, and these conditions have a great influence on teachers' behavior and beliefs, the researcher tries to find out how this relationship is constructed and the possibilities for the future during the transition.

A hallmark of a good qualitative case study is that it presents an in-depth understanding and description of the case. The themes or issues might be organized into a chronology by the researcher, analyzed across cases for similarities and differences among the cases, or presented as a theoretical model. This article reports a 3-year longitudinal case study of the transformation of the professional identities of four Chinese teachers during the first years of teaching in K-12 schools in China. The case study began with a detailed description of the differences between the imagined identity and practiced identity, and then the author explained that the social cognition representation theory is used as a conceptual framework. After all this information proposed, the author raised the two research questions: the first question is "What were the novice teacher's imagined identities like at the start of their teaching?" And the second question is "How did their imagined identities transform into practiced identities in the first years of teaching?" Then it comes to the participants chosen for this research. Four female participants - Ingrid, Carol, Fiona, and Aurora (pseudonyms), all 22 years of age when they became teachers - were supervised as fourth-year students in the teaching practicum of a BA program for ESOL teacher education at a normal university in Beijing.

The selection of how to approach the data analysis in a case study will differ. In this article, data were collected through the multiple sources of information, such as nonstructured interviews, observations and observation notes, participants' journals and observation sessions and its record materials. Then the author administered three rounds of data analysis. Firstly, the author reads and rereads all the data and entered a code. Secondly, existing codes' relationships were more rigorously studied and been classified into tentative categories and subcategories. Thirdly, the data analysis led to the reorganization of the tentative categorization by which the final conceptual framework was constructed.

Case studies often end with conclusions formed by the researcher about the overall meaning derived from the case(s). When it comes to the results and discussion, the author outlines how the four participants' imagined identities transformed into practiced identities in the novice stage. The author gives a very careful and detailed presentation about the discussion notes between the author and participants and give an analysis after each sample.

Finally, for the implications and conclusions, the findings suggest that novice teacher's cue-based or exemplar-based imagined identities may transform into rule-based or schema-based practiced identities as mediated by the influences of the institutional contexts of school and dynamic educational contexts and the institutional pressures seem to cause the imagined identities collapse or even be negatively replaced, but a teacher's perseverance and agency in keeping to his or her imagined identity may ultimately determine the positive evolution of that identity.

In this case study, the author follows Lincoln and Guba's (1985) case study structure—the problem, the context, the issues, and the "lessons learned". It also adds its own personal perspective by presenting information about the extent of our data collection and data analysis. The author identified the case for the study. This "case" was a bounded system, bounded by time (three-year of data collection) and four novice teachers that come from the same normal university of Beijing but teaching in different places during this research. One of the challenges inherent in qualitative case study development is that the researcher must identify the case. The case selected may be broad in scope or narrow in scope. The case study researcher must decide which bounded system to study, recognizing that several might be possible candidates for this selection and realizing that either the case itself or an issue, which a case or cases are selected to illustrate, is worthy of study. The researcher must consider whether to study a single case or multiple cases.

Selecting the case requires that the researcher establish a rationale for his or her purposeful sampling strategy for selecting the case and for gathering information about the case. Having enough information to present an in-depth picture of the case limits the value of some case studies. Deciding the "boundaries" of a case—how it might be constrained in terms of time, events, and processes—may be challenging. Some case studies may not have clean beginning and ending points, and the researcher will need to set boundaries that adequately surround the case.

5. CONCLUSION

A useful perspective to begin the process of differentiating among the five approaches is to assess the central purpose or focus of each approach. As shown in the figure, the focus of a narrative is on the life of an individual, and the focus of a phenomenology is on a concept or phenomenon and the "essence" of the lived experiences of persons about that phenomenon. In grounded theory, the aim is to develop a theory, whereas in ethnography, it is to describe a culture-sharing group. In a case study, a specific case is examined, often with the intent of examining an issue with the case illustrating the complexity of the issue. Turning to the five studies, the foci of the approaches become more evident.

As Richards concluded, great developments have taken place since the beginning of the new millennium: the broader field of qualitative research has been characterized by a less confrontational orientation, with attention shifting to practical issues and away from more theoretical debates. This is a welcome development in this field. From the samples above, it is reassuring to see a pragmatic approach gaining ground, as a result of which, it will be a great improvement of the development of researching methods in the applied linguistics area.

REFERENCES

- Allison, D. (2002). *Approaching English language research Singapore*: Singapore University Press.
- BENSON, P., CHIK, A., GAO, X.,HUANG, J.&WANG, W. (2009). Qualitative Research in Language Teaching and Learning Journals, 1997-2006. *The Modern Language Journal, 93*(1), pp.79-90.
- Block, D. (1996). Not so fast: Some thoughts on theory culling, relativism, accepted findings and the heart and soul of SLA. *Applied Linguistics, 17*, 63–83.
- Canagarajah, A.(2012). Teacher Development in a Global Profession: An Autoethnography. *TESOL Quarterly, 46*(2), pp.258-279.
- Christain, W.Chun(2014). Addressing Racialized Multicultural Discourses in an EAP Textbook: Working Toward a Critical Pedagogies Approach, *TESOL Quarterly,*109-421.
- Creswell, J.& Poth, C.(2017). *Qualitative inquiry & research design*. Los Angeles: Sage Publications, pp.53-54.
- Creswell, J.& Poth, C.(2017). *Qualitative inquiry & research design*. Los Angeles: Sage Publications, pp.83-120.
- Creswell, J., Hanson, W., Clark Plano, V. & Morales, A. (2007). Qualitative Research Designs. *The Counseling Psychologist, 35*(2), pp.236-264.
- Creswell, J.(1996). Research Design: Qualitative and Quantitative Approaches. John W. Creswell. *The Library Quarterly, 66*(2), pp.225-226.
- Davis, K. A.(1995). Qualitative theory and methods in applied linguistics research. *TESOL Quarterly, 29*,427–453.
- Egbert, J.(2007). Quality analysis of journals in TESOL and applied linguistics. *TESOL Quarterly 41.1*,157–171.
- Emerson, R. M., R. I. Fretz & L. L. Shaw (2001). Writing ethnographic fieldnotes. In Atkinson et al.(eds.), 352–368.
- En.wikipedia.org. (2017). *TESOL Quarterly*. [online] Availableat: https://en.wikipedia.org/wiki/TESOL_Quarterly [Accessed 29 Dec. 2017].
- Gollin-Kies, S. (2014). Methods reported in ESP research articles: A comparative survey of two leading journals. *English for Specific Purposes, 36*, pp.27-34.
- Gubrium J. F. & J. A. Holstein (eds.) (2002). *Handbook of interview research: Context and method*. Thousand Oaks, CA: Sage.
- Jones, J. F.(2004). The many benefits of a research component in English language teacher education:A case study. *Prospect 19.2*, 25–38.
- Lazaraton, A.(1995). Qualitative research in applied linguistics : A progress report. *TESOL Quarterly, 29*,455–472.
- Lazaraton, A. (2002). Quantitative and qualitative approaches to discourse analysis. *Annual Review of Applied Linguistics, 22*, 32–51.
- Lazaraton, A. (2003b). Incidental displays of cultural knowledge in the nonnative-English-speaking teacher's classroom. *TESOL Quarterly 37.2*, 213–245.
- Leibowitz, B.(2005). Learning in an additional language in a multilingual society: A South African case study on university-level writing. *TESOL Quarterly 39.4*, 661–681.
- Richards, K.(2003). *Qualitative inquiry in TESOL*. London : Palgrave Macmillan.
- Richards, K.(2009). Trends in qualitative research in language teaching since 2000. *Language Teaching, 42*(02), p.147.

- Stapleton, P.& Shao, Q. (2017). Research in language teaching over two decades: A retrospective of the first 20 volumes of Language Teaching Research. *Language Teaching Research*, p.136216881668892.
- Wright, M.& Cook, V.(1994). Second Language Learning and Language Teaching. *Language, 70*(1), p.198.
- Xu, H.(2012).Imagined Community Falling Apart: A Case Study on the Transformation of Professional Identities of Novice ESOL Teachers in China. *TESOL Quarterly, 46*(3), pp.568-578.

G200

ASSOCIATION

www.g200youthforum.org

G200 Association
14 rue du Rhone
Geneva 1204, Switzerland

ISBN 978-1-5108-6183-1

9 781510 861831

CURRAN ASSOCIATES INC.
proceedings.com

International Conference on Applied Mathematics, Computational Science and Systems Engineering (AMCSE 2017)

ITM Web of Conferences Volume 16 (2018)

Athens, Greece
6 - 8 October 2017

edp sciences